Student Learning Guide

to accompany

BIOSPHERE 2000:
Protecting Our Global Environment
third edition

Donald G. Kaufman
Miami University

Cecilia M. Franz
Miami University

with contributions by
Cheryl Puterbaugh
and numerous Miami University students

KENDALL/HUNT PUBLISHING COMPANY
4050 Westmark Drive   Dubuque, Iowa 52002

All unit openers, Susan Friedmann

Grateful acknowledgment is hereby made to HarperCollins Publishers
to reprint illustrations from *Biosphere 2000,* first edition.

Copyright © 1999 by Donald G. Kaufman and Cecilia M. Franz

ISBN 0-7872-6470-9

Kendall/Hunt Publishing Company has the exclusive rights to reproduce this work,
to prepare derivative works from this work, to publicly distribute this work,
to publicly perform this work and to publicly display this work.

All rights reserved. No part of this publication may be reproduced,
stored in a retrieval system, or transmitted, in any form or by any
means, electronic, mechanical, photocopying, recording, or otherwise,
without the prior written permission of Kendall/Hunt Publishing Company.

Printed in the United States of America

10   9   8   7   6   5   4   3   2   1

Table of Contents

Preface and Acknowledgments v

Introduction: A Learning Process 1

Unit I: The Biosphere and Environmental Studies

Chapter 1. An Overview of Environmental Problems 6
Chapter 2. Science and Environmental Studies 15

Unit II: An Environmental Foundation: Ecological Principles and Applications

Chapter 3. Ecosystem Structure 24
Chapter 4. Ecosystem Function 32
Chapter 5. Ecosystem Development and Dynamic Equilibrium 40
Chapter 6. Ecosystem Degradation 49
Chapter 7. Applying Ecological Principles 56

Unit III: An Environmental Imperative: Balancing Population, Food, and Energy

Chapter 8. Human Population Dynamics 64
Chapter 9. Managing Human Population Growth 75
Chapter 10. Food Resources, Hunger, and Poverty 85
Chapter 11. Energy Issues 96
Chapter 12. Energy: Fossil Fuels 103
Chapter 13. Energy: Alternative Sources 111

Unit IV: An Environmental Necessity: Protecting Biospheric Components

Chapter 14. Air Resources 123
Chapter 15. Water Resources 132
Chapter 16. Soil Resources 142
Chapter 17. Biological Resources 150

Unit V: An Environmental Pandora's Box: Managing the Materials and Products of Human Societies

Chapter 18. Mineral Resources 160

Student Learning Guide

Chapter 19. Nuclear Resources 169

Chapter 20. Toxic and Hazardous Substances 177

Chapter 21. Unrealized Resources: Waste Minimization and Resource Recovery 186

Unit VI: An Environmental Heritage: Preserving Threatened Resources

Chapter 22. The Public Lands 195

Chapter 23. Wilderness 203

Chapter 24. Cultural Resources 210

Unit VII: An Environmental Legacy: Shaping Human Impacts on the Biosphere

Chapter 25. Religion and Ethics 218

Chapter 26. Economics and Politics 225

Chapter 27. Law and Dispute Resolution 236

Chapter 28. Environmental Education 244

Special Supplement: Guidelines for Research 251

Answer Section 260

Preface and Acknowledgments

Those readers who have been with us since the first edition of *Biosphere 2000: Protecting Our Global Environment* will have noticed a pattern over the years. With each edition, we rely for help on talented associates, many of them undergraduates. Such was the case with this third edition of the *Student Learning Guide to accompany Biosphere 2000*. We owe a great deal of thanks to Miami students Tony Nardini and Stefanie Brown, who labored many hours to produce a helpful, effective *Guide*. Thanks, too, to our associate, Lisa Rosenberger, who oversaw the revision process and who took great care to ensure that this *Guide* reflects the extensive changes in the third edition of the textbook. Lisa was ably assisted throughout the revision by Miami student Kimberly Bruce, whose efforts we much appreciate. We also are grateful to Carole Katz, Art Director for *Biosphere 2000: Protecting Our Global Environment,* for her help in the design and layout of this *Guide*.

As you complete your course, we hope you'll find this third edition of the *Student Learning Guide to accompany Biosphere 2000* helpful. We believe it can be a great learning tool, but as with any tool, its worth is proven with use. So, please use this *Guide* — and remember, it was written with you, the student, in mind.

Donald G. Kaufman and Cecilia M. Franz, June 1999

Introduction

A Learning Process

You've probably already been given lots of tips and techniques for studying. You've been told to schedule your time, get plenty of rest, eat the right foods, and read in a quiet environment. You know four different methods for taking notes in class and several more techniques for outlining chapters. You can study alone, or with a partner, or in a group.

But these things affect only the environmental and mechanical aspects of what's going on. They never get to the heart of the matter, which isn't how you study, but how you learn. All those study techniques can be useful, and certainly they're full of common sense, but they'll work only if you pair them with a real effort to learn. That's why this book isn't a *study* guide; it's a *learning* guide.

The *Student Learning Guide* is set up so that you can follow a step-by-step technique for learning. There isn't anything really original about this technique; it's just some tried-and-true practices plus some common sense ideas about how adults learn. Those practices are embodied in an old Chinese proverb:

I hear and I forget;

I see and I remember;

I do and I understand.

It happens that what the Chinese and others have known intuitively for a long time has recently begun to be confirmed by science and statistics. For example, one study concluded that we remember

10% of what we read

20% of what we hear

30% of what we see

50% of what we see and hear

80% of what we say

90% of what we say as we act

According to these findings, if you do nothing more than attend class and read the assigned chapters you'll remember about half of the class content!

Five Steps to Learning: Making the Most of Your Senses

The *Student Learning Guide* is set up so that you can use all of your senses to help you learn. In conjunction with the readings in the textbook *Biosphere 2000* and your instructor's in-class lectures, the exercises and activities in the *Guide* will help you follow a learning path that looks something like this:

- Understand the basic concepts.
- Add to those concepts the associated facts.
- Relate the concepts and facts to each other, and translate them into meaningful, useful knowledge.

The *Student Learning Guide* has taken those three basic tasks and turned them into five steps that are the foundation of a learning technique. Steps 1 & 2 correspond with the task of understanding the basic idea. Step 3 corresponds with the task of adding facts and figures to the idea. Steps 4 & 5 enable you to translate the information into personal and useful knowledge. We suggest that you follow this technique as you learn about the environment through reading *Biosphere 2000: Protecting Our Global Environment*.

 ## Step 1: Read for Chapter Objectives

Study principles

1. Know what to expect from the chapter.
2. Think about the objectives as you read.

Before you read the chapter, you should already know the most important points that will be made. What do the authors want you to know? The Chapter Outline and the Learning Objectives will indicate the things that are important for you to know. For example, the first major section in Chapter Two asks "What Is Science?" It's a safe bet that in Chapter Two, it will be important to gain a solid understanding of science, its strengths and limitations.

 ## Step 2: Identify the Important Concepts

Study Principle:

1. Grasp the major concepts first.

Identifying the important concepts means making a mental note of specific concepts as the appear and are elaborated on. The Key Concepts of each chapter can be identified in the chapter summary, or you can highlight in the text as you read. These concepts provide a framework for organizing the facts you will learn in Step 3.

Step 3: Master the Facts

Study Principles:

1. Read the fact aloud or picture it in your mind.

2. Apply it to a major concept.

3. Say or write it in your own words.

The important facts you need to master in Step 3 are things like terms, numbers, names and places. Your knowledge of the facts is tested with Multiple Choice, True/False, Fill in the Blank, and Short Answer questions. The Key Terms are also listed for each chapter.

You learn facts by figuring out some way to relate them to yourself or something you already know, by explaining them to someone else, by picturing them vividly in your mind, by reciting them, or by writing them in your own words. As you encounter each fact, you should associate it with a concept you identified in Step 2. Think about the fact — What does it really mean? What concept does it apply to and how does it fit in? Imagine what it looks like. For things that are alike, make sure you can tell what's different about them. For things that are parts of a whole, make sure you can tell how they fit together — and break down again. For things that are listed together, make sure you know how they relate to each other.

Examples are helpful at this point. It's easier to understand and remember an example drawn from a real-life situation than it is to memorize a dictionary definition. Examples also provide you with a mental framework, so by remembering the example, you remember the facts. When you get to Step 4, you'll use these facts to support or illustrate your own discussion of these concepts.

Step 4: Analyze, Compare and Apply

Study Principles:

1. Practice relating the facts and concepts to each other.

2. Use examples to illustrate the concepts.

3. Back up your essays with the facts.

In this step, you perform the higher-order learning tasks that are often tested with essay questions and research paper assignments. The *Student Learning Guide* uses Related Concepts and Thought Questions to provide you with the opportunity to discuss your new knowledge.

 Step 5: Put It into Practice

Study Principles:

1. Relate what you've learned to your own experience.

2. Explain what you've learned to someone else.

Without a personal experience which imparts significance to the concepts and facts that you learn, you can quickly forget the information. In fact, research has shown that you'll usually forget it in 48 hours. Step 5 is the final, important step, in which you use the information you have learned and make it part of your experience.

One of the best ways to put your learning into practice and make sure you've really got it is to teach it to someone else. When you know you have to teach something, you learn it better, because you can't get by with a partial understanding. Try to arrange with a friend or classmate to teach them the material. If you can't do that, make use of the Suggested Activities sections in the *Student Learning Guide*. These activities are intended to give you a real-life framework for internalizing what you've learned.

To make all this work for you, you can follow these steps as you learn the material in *Biosphere 2000*. As you go through each of the exercises in this *Guide*, you'll see one of the five icons that represent these steps at the beginning of each new section. The icon is there to remind you where you are in the learning process and what you should be doing.

We have also included interesting pieces of information in "Did You Know...?" boxes. These bits of not-so-trivial trivia help to bring some real-world perspectives to your academic study. You can also use them to impress your friends with your environmental knowledge! Additionally, each chapter of the *Guide* includes an "Environmental Success Story," a feature that highlights positive environmental news.

Here is one final piece of advice: Feel free to write in this learning guide wherever you want. You will notice that we have left space for you to write in the outline and around the questions. Take advantage of this space to add information you think is important or to make any other notes that will be helpful to you. This *Guide* was written to help you learn — don't hesitate to make it your own.

UNIT I

The Biosphere and Environmental Studies

Chapter 1
An Overview of Environmental Problems

 Chapter Outline

I. **What Is the Biosphere?**

   A. Definition

      1. Thin layer of rock, air, water, and soil surrounding Earth

      2. Contains conditions to support life

   B. Living and Nonliving Components

      1. Abiotic (nonliving)

      2. Biotic (living)

   C. Characteristics of Living Organisms

      1. Live at the Expense of Their Environment

      2. Have a Cellular Structure

      3. Exhibit Movement

      4. Grow

      5. Reproduce

      6. Respond to Stimuli

      7. Evolve and Adapt

II. **How Did the Biosphere Develop?**

   A. Big Bang Theory

      1. Universe arose from hot point called a singularity.

      2. Singularity exploded and space began to expand.

      3. As expanding universe cooled, particles and matter formed.

      4. As gravity caused matter to form clumps, galaxies formed.

   B. Life Developed on Earth

      1. Life developed 3.5 to 4 billion years ago.

2. Most probably the result of chemical and physical processes on the planet surface

3. "Primordial Soup"

C. Evolution Proceeded on Young Planet

1. Result of organisms' adaptations to environmental changes

2. Driving forces

a. Mutations

b. Natural selection

III. What Is the State of the Biosphere?

A. Indications of Trouble

1. Despoiled Beaches

2. Hole in the Ozone

3. Severe Drought

4. Devastating Hurricanes

IV. What Are the Three Root Causes of Environmental Problems?

A. Population Growth

1. Exponential Growth

2. Linear Growth

B. Abuse of Resources and Natural Systems

1. Renewable Resources

2. Nonrenewable Resources

C. Pollution

1. Natural Sources

2. Cultural Sources

V. Why Do Uncontrolled Population Growth, Resource Abuse, and Pollution Occur?

A. Worldview

1. Person's/society's way of perceiving reality

2. Comprised of attitudes, values, and beliefs

3. Reflected in and transmitted through culture

B. Dominant Western Worldview a Product of:

1. Judeo-Christian Beliefs
  2. Reliance on Science
  3. Rise of Capitalism and Democracy
  4. Industrialization
 C. Western Worldview Emphasis
  1. Exploitation of Resources
  2. Accumulation of Wealth
  3. Faith in Science
  4. Belief in the Inherent Right of the Individual

VI. What Is the Environmental Movement?
 A. Public Awareness
  1. Crystallizes in the 1960s
  2. Aldo Leopold
  3. Rachel Carson
 B. Laws
 C. Politics

VII. How Has the Environmental Movement Caused Us to Reassess Our Relationship with Nature?
 A. Stewardship Ethic
 B. Anthropocentrism
 C. Biocentrism

Learning Objectives

After learning the material in Chapter 1, you should be able to:

1. Describe the biosphere and explain how scientists believe it evolved.
2. Summarize the current state of the biosphere and identify the three root causes of environmental problems.
3. Distinguish between linear and exponential growth and explain the significance of exponential human population growth.
4. Explain how environmental problems and the human activities that cause them are related to cultural attitudes, values, and beliefs.
5. Distinguish between a biocentric and an anthropocentric worldview.

DID YOU KNOW...?

In the second it takes to snap your fingers, five people will be born and two people will die — a net gain of three people a second. At this rate, 259,200 people are born each day.

 ## Key Concepts

Read this summary of Chapter 1 and identify the important concepts discussed in the chapter.

The biosphere is the thin layer of air, water, soil, and rock that surrounds the Earth and contains the conditions to support life. Scientists believe that the universe originated from an infinitely dense, infinitely hot point, called a singularity, in a big bang some 13 to 20 billion years ago. The Earth formed about 4.6 billion years ago from interstellar dust and gas. Life on Earth is believed to have begun between 3.5 and 4 billion years ago, after the planet's surface cooled. Over the millennia, the Earth's living organisms evolved and diverged, adapting to changes in the nonliving environment. The process of evolution is driven by genetic mutations and natural selection. Mutations, random changes within the genetic material of an individual that can be passed to that individual's offspring, are often harmful or result in changes that are not useful to the organism. Sometimes, however, a mutation causes an organism to differ in such a way that it is better suited to its environment than are other members of the population. Because it is more likely to survive and reproduce, passing on its genetic characteristics to its young, it is said to be "naturally selected for." Natural selection thus enhances an organism's chances for successful reproduction.

While change has been a constant characteristic of the Earth throughout its history, humans have altered the planet in unprecedented ways. We have been able to modify our environment significantly through technology and social institutions, but we cannot control natural processes. As the current state of the biosphere illustrates, *our Earth environment sets limits on all creatures*. Natural catastrophes such as droughts and hurricanes, which may or may not be exacerbated by human actions, and worldwide problems such as global warming and stratospheric ozone depletion, indicate that we may have approached the Earth's limits.

The three root causes of environmental problems are population growth, abuse of resources and natural systems, and pollution. Global population growth is exponential; increased growth in many parts of the world is stressing or exceeding the productive or supportive ability of local natural systems. In slow growth areas such as the United States, per capita consumption is so high that a single individual has a far greater impact on the environment than does an individual in a fast growth (generally poor) country. Resource abuse occurs when renewable resources are used in such a way that they cannot be regenerated; they are exploited to such an extent that they become depleted. Satisfying wants rather than needs can lead to resource abuse, as can technologies that harvest or exploit resources so completely or so efficiently that the resources cannot renew themselves. Pollution may be local, regional, or global in nature, and it may originate from natural or cultural sources. It is a mistake to suppose that pollution is a contemporary phenomenon; humans (and other organisms) have always generated wastes. However, as long as populations were small, natural systems could degrade and recycle wastes. Unfortunately, more populous contemporary societies produce synthetic wastes and wastes in such large volumes that natural systems are often unable to recycle them.

Environmental problems arise from the interaction of natural and cultural systems. Accordingly, to solve environmental problems, individuals and societies must address the underlying cultural factors — attitudes, values, and beliefs — that cause the problems. Part of that process must be an examination of one's worldview, or way of looking at reality, which includes beliefs about the relationship of humans with the natural world. It is also necessary to examine how we define "nature" and "environment." Do we consider humans part of nature? Do we consider humans part of the environment? Why or why not? What are the implications of our definitions of these terms?

At the dawn of the new millennium, it is becoming increasingly clear that protecting our global environment requires us to develop and encourage aspirations within the limits imposed by the environment. Ecological balance is the ultimate relationship humans must learn to maintain with nature. *No matter how far removed we become from direct contact with nature, we remain a part of it.* If our species is to survive and prosper, this contact must be nurtured.

 ## Key Terms

abiota	anthropocentrism
big bang theory	biocentrism
biosphere	biota
biotic	environment
evolution	exponential growth
frontier mentality	land ethic
linear growth	mutation
natural selection	nature
net primary productivity (NPP)	nonrenewable resource
perpetual resource	pollutant
renewable resource	resource
sense of the Earth	speciation
stewardship ethic	worldview

Environmental Success Story

By treating kids as problem solvers, KIDS as Planners, a program of KIDS (Kids Involved Doing Service) Consortium, has demonstrated that children can make substantial contributions to their communities. The program brings together kindergarten through high school students to help communities solve real-world environmental problems. It also shows teachers how they and their students can team up with local agencies and businesses to help monitor rivers, reduce nonpoint-source pollution, develop land use management plans, and promote the use of alternate transportation. Over the past four years, more than 5,000 students, 350 teachers, and 50 schools nationally have implemented the KIDS model.

Source: KIDS Consortium, Inc. "Eighth Annual National Awards for Environmental Sustainability Winners: KIDS as Planners." *Renew America* (22 April 1998). Available: http://solstice.crest.org/sustainable/renew_america/winner98.html. 24 March 1999.

True/False

1. The biosphere is composed of two distinct and entirely unrelated components, the abiota and the biota. T F

2. Living organisms evolve and adapt in response to changes in the environment because of an innate desire to survive. T F

3. Earth underwent significant change from the time of its formation until the appearance of life, approximately 3.5 to 4 billion years ago, but the physical environment has remained relatively constant since then. T F

4. Resource consumption in affluent countries has a serious impact on the planet's natural systems. T F

5. Religion is not a significant factor in the development of a person's or society's worldview. T F

Fill in the Blank

1. According to the United Nations, of the nearly six billion people now living on the Earth, only _____ have adequate food, housing, and safe drinking water.

2. A substance that adversely affects the quality of the Earth's environment is a(n) _____.

3. The belief that buildings and other things produced by humans are not natural is part of a(n) _____ view.

4. The worldview held by St. Francis of Assisi and other Christians saw humans as _____ of the Earth.

5. Beginning at the end of the eighteenth century, _____ and _____ caused people to become separated from daily contact with the land, and they began to lose their sense of the Earth.

Multiple Choice

Choose the best answer.

1. Which of the following terms describe opposite conceptions of the human in nature?

    A. biotic and abiotic

    B. biocentric and anthropocentric

    C. cultural and primitive

    D. consumerism and environmentalism

2. The biosphere

    A. is made up of the interacting regions of the atmosphere, hydrosphere, and lithosphere.

    B. includes any place where life can be found.

    C. is relatively small compared to the Earth's total mass.

    D. All of the above are true.

3. The belief that humans are subject to all natural laws is characteristic of the _____ viewpoint.

    A. Spaceship Earth

    B. biotic

    C. anthropocentric

    D. biocentric

4. The three root causes of environmental problems are

    A. environmental degradation, decreasing species diversity, and population growth.

    B. consumerism, resource abuse, and increased consumption per capita.

    C. population growth, resource abuse, and pollution.

    D. industrialization, pollution, and population growth.

5. A key factor that contributed to the environmental movement of the 1960s and 1970s was

   A. writings, such as *Silent Spring* by Rachel Carson, that warned of environmental degradation.

   B. fear of global warming.

   C. a growing belief in anthropocentrism.

   D. All of the above are true.

Short Answer

1. What is the Spaceship Earth analogy?
2. Define "anthropocentric" and "biocentric" in terms of a person's definition of what is natural.
3. What characteristics are shared, in general, by all living organisms?
4. What are renewable resources? What are nonrenewable resources?
5. What is net primary productivity (NPP)?

Thought Questions

Develop a complete answer for each of the following.

1. What is meant by the following statement? *Our Earth environment sets limits on all creatures.* What viewpoint does this represent?
2. How do the abiota and biota interact with one another to influence the biosphere?
3. What is the prevailing scientific theory of how life on Earth began? How did evolutionary processes produce the species on Earth today?
4. How are the three root causes of environmental problems related?
5. Explain the difference between linear and exponential growth and why this is important in terms of human population growth.
6. What are the implications of the population growth race for people in developing countries? For nonhuman species? For nonrenewable resources?
7. What is the Western worldview and how has it affected the environment? What are the major factors that shaped the Western worldview?
8. What social and environmental conditions contributed to the environmental movement of the 1960s and 1970s? How successful was this movement?

▪ Related Concepts

Describe the relationship. (There may be more than one.)

BETWEEN...	AND...
exponential growth	population
cultural systems	biosphere
ozone depletion	acid precipitation
Aldo Leopold	land ethic
frontier mentality	Native American beliefs

Did You Know...?

The average American throws away about 3.2 pounds of trash a day — that's 99 pounds a month and over 1,100 pounds a year.

 ## Suggested Activities

1. Exponential growth can be a tricky concept to grasp. To get a better understanding of how population numbers increase so dramatically, use your calculator to carry out the example given in Chapter 1. If every day for a month you are given twice the amount you received the day before, starting with one penny, how much would you receive on the 30th day?

2. Research an environmental problem that has affected you or someone you know. What factors led to its occurrence? What are the effects? What is being done to solve it?

3. In your opinion, what is the most serious environmental problem facing the world today? Think about why you feel this way — what beliefs and values do you hold that led you to conclude this? Write an essay about your answer.

4. Take some time to think about your worldview and write a statement describing it. Think about the consequences of your beliefs, attitudes, and values. What sort of Earth do you want to leave for your children and for all future generations to inherit? What will it take to make that legacy possible?

5. Start a microcosm and observe it for a few weeks or months. Fill a clean jar with water from a pond, lake, marsh or tidal pool, seal the jar so that it is airtight, and place it on a windowsill where it will receive indirect sunlight. (Too much direct sunlight will overheat the jar and kill the organisms inside.)

Chapter 2
Science and Environmental Studies

 Chapter Outline

I. **What Is Science?**

   A. Scientific Inquiry — A Way of Knowing about the Natural World

      1. Observation

      2. Hypothesis

      3. Controlled Experiment

      4. Results

      5. Significance

   B. Strengths of Scientific Inquiry

   C. Limitations of Scientific Inquiry

II. **What Is Ecology?**

   A. Definition

      1. Scientific study of structure, function, and behavior of natural systems that comprise the biosphere

      2. Study of relationships of organisms with each other and with the environment

   B. An Attempt to Understand Ecosystems as Functioning Systems

III. **What Is Environmental Science?**

   A. Definition

      1. Study of the human impact on the physical and biological environment of an organism

      2. Can encompass the social and cultural aspects of the environment

IV. **What Is Environmental Studies?**

   A. Definition

      1. Interdisciplinary field that attempts to solve the problems caused by the interaction of natural and cultural systems

      2. Differs significantly from science — takes into account human values

      3. Includes contributions of both scientists and non-scientists

V. What Methods Can We Use To Solve Environmental Problems?

A. Environmental Problem Solving

1. Utilizes the five-step environmental problem-solving model

   a. Identify and diagnose the problem.

   b. Set goals and objectives.

   c. Design and conduct a study.

   d. Propose alternative solutions.

   e. Implement, monitor, and reevaluate the chosen solution.

2. Advantages

   a. Interdisciplinary

   b. Produces a workable solution that can be monitored and adapted as needed

3. Disadvantages

   a. Reactive method of problem solving

   b. Damage must occur before problem is recognized.

B. Environmental Activism

C. Litigation

VI. How Can We Minimize Environmental Problems?

A. Stewardship Ethic

B. Environmentally Sound Management

1. Premise: Any resource can be managed in an environmentally sound way.

2. Management plan minimizes or prevents environmental degradation.

3. Similar to environmental problem solving, but incorporates a stewardship ethic

VII. How Can We Implement Environmentally Sound Management?

A. Proactive Management of Resources

B. Management Constructed Through Use of the Following Criteria:

1. Stewardship Ethic

2. Biocentric Worldview

3. Understanding of Natural System Dynamics

4. Environmental Education

5. Interdisciplinary Planning
6. Data Based on Sound Natural and Social System Research
7. Sociocultural Considerations
8. Knowledge of Political Systems
9. Sound Economic Analysis
10. Maximum Public Participation

Learning Objectives

After learning the material in Chapter 2, you should be able to:

1. Distinguish between science, ecology, environmental science, and environmental studies.
2. Describe the process of scientific inquiry.
3. Explain how knowledge from the sciences, social sciences, and arts and humanities contributes to solving environmental problems.
4. Identify three different methods for solving environmental problems.
5. List the five steps of the problem-solving model for environmental studies and describe what each step entails.
6. Describe the major characteristics of environmentally sound management and explain why it is preferable to environmental problem solving.

Did You Know...?

According to the Federal Reserve Board, chemical production in the United States increased 1,326 percent between 1947 and 1994.

 ## Key Concepts

Read this summary of Chapter 2 and identify the important concepts discussed in the chapter.

The term "science" refers to a body of systematized knowledge about nature and the physical world. Scientific inquiry — a process of making observations, developing hypotheses, and conducting controlled experiments to test those hypotheses — is used to acquire this knowledge. Ecology is a scientific discipline that uses the scientific method to study the structure, function, and behavior of the natural systems that comprise the biosphere.

Environmental studies differs from ecology and other natural sciences because it takes into account human values and culture. Environmental problems contain social, political, and economic dimensions; to solve them, we must draw upon knowledge in numerous and diverse areas. There are no easy solutions to these problems. Our challenge in solving them is to understand the underlying scientific principles and the consequences of interrupting natural processes through the long-term effects of our actions.

The five steps in our environmental problem-solving method include: (1) identify and diagnose the problem; (2) set goals and objectives; (3) design and conduct a study; (4) propose alternative solutions; and (5) implement, monitor and reevaluate the best solution. Other methods that can be used to solve environmental problems are litigation (the initiation of a lawsuit) and environmental activism.

Environmental problems can be minimized or avoided through environmentally sound management, which acknowledges that all forms of life on Earth have value; opposes uncontrolled resource exploitation; promotes the wise use of resources; and minimizes waste and environmental damage. Environmentally sound management is based on several considerations, including a stewardship ethic, a biocentric worldview, environmental education, interdisciplinary planning, and maximum public participation. Since no one can predict what future generations will deem of value, the most important thing we can do is ensure a maximum of choices for all generations.

 ## Key Terms

ecology

environmental science

environmental studies

science

scientific method

environmental activism

environmentally sound management

litigation

scientific inquiry

Environmental Success Story

The Tulsa Postal Service facility found they had 250 gallons of excess paint. Instead of disposing of the paint, the facility donated it to the local Tulsa Habitat for Humanity group. This simple decision not only benefited the Habitat for Humanity group by providing it with essential building materials, but also benefited the environment by reducing waste.

Source: "Environmental Success Story, Oklahoma District." *United States Postal Service Website.* Available: http://www.usps.com/environ/textmirr/webpages/oklahoma.htm. 16 February 1999.

True/False

1. Given sufficient funds for research and development of advanced technologies, science will *always* find effective solutions to environmental problems that are acceptable to all parties. T F

2. The most difficult, and the most easily overlooked, step of the problem-solving model is the identification and diagnosis of the problem. T F

3. Litigation is the preferred method for solving environmental problems because it offers the best outcome for the environment (that is, it offers the best protection for the environment and natural systems). T F

4. Environmental problem solving is relatively inexpensive and proactive while environmental management is costly and reactive. T F

5. Environmental science is synonymous with ecology. T F

Fill in the Blank

1. _____ is the study of the structure, function, and behavior of the natural systems that comprise the biosphere.

2. Observation, hypothesis development, and experimentation together comprise _____.

3. A(n) _____ is a self-sustaining community of organisms interacting with one another and with the physical environment within a given geographic area.

4. _____ is an adversarial process in which opposing sides are represented by legal attorneys; it can be used to address environmental problems.

5. Environmentally sound management differs from environmental problem solving because it incorporates a(n) _____.

Multiple Choice

Choose the best answer.

1. _____ attempts to produce the least environmentally disruptive decisions through a management plan that prevents or minimizes environmental degradation.

    A. Litigation
    B. Environmental problem solving
    C. Environmentally sound management
    D. Environmental activism

2. A(n) _____ is designed to compare two situations that differ in a single variable.

   A. environmental problem-solving model

   B. lawsuit

   C. controlled scientific experiment

   D. environmentally sound resource management model

3. The five steps of the problem solving model, in correct order, are:

   A. set goals and objectives; identify and diagnose the problem; design and conduct a study; propose alternative solutions; implement, monitor, and evaluate the chosen solution.

   B. identify and diagnose the problem; set goals and objectives; design and conduct a study; propose alternative solutions; implement, monitor, and evaluate the chose solution.

   C. design and conduct a study; identify and diagnose the problem; set goals and objectives; propose alternative solutions; implement, monitor, and evaluate the chosen solution.

   D. propose alternative solutions; set goals and objectives; identify and diagnose the problem; design and conduct a study; implement, monitor and evaluate the chosen solution.

4. Which of the following is NOT a method for resolving difficult environmental issues and situations?

   A. environmental problem solving

   B. litigation

   C. environmentally sound management

   D. environmental activism

5. Environmental science differs from ecology in that it

   A. considers human values.

   B. defines a process for solving environmental problems.

   C. must sometimes take action on a problem or issue before all data are collected.

   D. All of the above are true.

Short Answer

1. Describe the process of scientific inquiry.
2. What are two ways that scientists ensure their research is free of bias or coercion?

3. What is an ecosystem?
4. Explain briefly how each of the following may play a role in environmental problem solving: ecology, theology, education, art, communications, history, political science, and economics.
5. What are the five steps of the environmental problem-solving model?

Thought Questions

Develop a complete answer for each of the following.

1. What are the differences between science, ecology, environmental science, and environmental studies?
2. What are the strengths and limitations of scientific inquiry?
3. What are the differences between the scientific method and environmental problem solving? What are the similarities? When is one more appropriate than the other?
4. Explain the three methods of solving environmental problems (five-step model, litigation, activism) and discuss why environmentally sound management is preferable to all three.
5. Describe the members of a typical environmental problem-solving team. Explain why interdisciplinary planning is essential to the success of the environmental problem-solving method.

▫▪■ Related Concepts

Describe the relationship. (There may be more than one.)

BETWEEN...	AND...
hypothesis	principle
ethics	science
value judgments	environmental problems
litigation	environmental problem-solving model
environmental problem solving	environmentally sound management
scientific method	environmental problem solving

Did You Know...?

Some paper can be recycled seven times or more!

Suggested Activities

1. Write a letter to the editor of your local newspaper, encouraging citizens to adopt a stewardship ethic toward the environment.

2. Consider your own intended major or profession. How might you contribute to environmental problem solving and environmentally sound management?

3. Imagine you have been asked to solve an environmental problem. Describe the interdisciplinary team you would need to follow the environmental problem-solving model, and explain each of the your choices.

4. Develop your own stewardship ethic by nurturing your sense of the Earth. You might tend a garden, visit a national park or wilderness area, or search for wildflowers along roadsides or in vacant lots. Capture your thoughts and experiences through writing, art, or some other means of expression.

UNIT II

An Environmental Foundation:
Ecological Principles and Applications

Chapter 3
Ecosystem Structure

 Chapter Outline

I. **What Are the Levels of Ecological Study?**

   A. Individual

   B. Species

   C. Population

   D. Community

   E. Ecosystem

   F. Biome

   G. Biosphere

II. **What Are the Components of An Ecosystem?**

   A. Abiotic Components

   1. Energy

      a. First law of thermodynamics

      b. Second law of thermodynamics

      c. Entropy

   2. Matter

      a. Law of conservation of matter

      b. Macronutrients

      c. Micronutrients

   B. Biotic Components

   1. Producers (Autotrophs)

      a. Phototrophs

      b. Chemotrophs

   2. Consumers (Heterotrophs)

      a. Macroconsumers

      b. Microconsumers

III. What Determines the Structure of Ecosystems?

    A. Abiotic Limiting Factors

        1. Temperature

        2. Light

        3. Oxygen

    B. Biotic Limiting Factors

        1. Prey

        2. Predator

Learning Objectives

After learning the material in Chapter 3, you should be able to:

1. Identify the levels of ecological study and briefly define each.
2. Define ecosystem and list the two major components of any ecosystem.
3. Define producers and consumers and give examples of each.
4. Describe the roles played by producers, consumers, and decomposers in ecosystems and discuss the interactions among them.
5. Identify three important limiting factors and give an example of how each helps to regulate the structure of an ecosystem.

Did You Know...?

Every gram of compost contains one billion organisms.

Key Concepts

Read this summary of Chapter 3 and identify the important concepts discussed in the chapter.

Ecologists study the natural world at many levels: individual, species, population, community, ecosystem, biome, and biosphere. An individual is a single member of a species. A species includes all organisms that are capable of breeding to produce viable, fertile offspring. Individuals of a particular species that live in the same geographic area comprise populations; populations have measurable group characteristics such as birth rates, death rates, seed dispersal rates, and germination rates. The place where the individual organism or population lives is its habitat. All of the populations of organisms that live and interact with one another in a given area at a given time are collectively known as a community. A community and its interactions with the physical environment comprise an ecosystem. A biome is a grouping of many terrestrial ecosystems in a specific, large geographic area identified by a dominant vegetation type. The union of all terrestrial and aquatic ecosystems — and the largest system of life-physical interactions on Earth — is called the biosphere.

Ecosystems are life-perpetuating systems having both biotic and abiotic components. The abiotic component consists of energy, matter, and other factors such as temperature and water. The Earth is an open system for energy; the constant flow of solar radiation supplies nearly all the energy that makes life on Earth possible. (A very small portion of energy is internal — energy trapped within the Earth at its formation. This energy makes its way to the surface through volcanoes, deep sea vents, geysers, and hot springs.) Less than one percent of incoming solar radiation is captured by green plants through the process of photosynthesis; most is either reflected by the Earth's cloud cover or is reradiated from the planet's surface back into space. Energy can neither be created nor destroyed but may be changed in form and may be moved from place to place, a principle known as the first law of energy, or first law of thermodynamics. The second law of energy, or second law of thermodynamics, states that with each change in form, some energy is degraded to a less useful form and given off to the surroundings, usually as heat. Entropy refers to the tendency of natural systems toward dispersal or randomness.

Matter is anything that has mass and takes up space. The Earth is essentially a closed system for matter, with the exception of the small amount of matter that enters the Earth's atmosphere via meteors and meteorites. The law of the conservation of matter states that during a physical or chemical change, matter is neither created nor destroyed, but it may be moved from place to place. An important application of the law of the conservation of matter is that materials are continually cycled and recycled through ecosystems.

The biotic component of an ecosystem is composed of autotrophs (producers) and heterotrophs (consumers). Most producers are phototrophs, receiving energy from the sun. Consumers include macroconsumers such as herbivores, carnivores, omnivores, scavengers, and detritivores. Microconsumers, or decomposers, play a vital role in reducing complex organic matter to inorganic matter and returning nutrients to the environment.

Both abiotic and biotic factors affect the structure of an ecosystem. Limiting factors such as temperature, light, and available nutrients are abiotic regulators. They and other abiotic regulators form a complex set of interactions that limit the activities of individual organisms, populations, and communities. Too little, or too much, of a particular limiting factor may affect the organisms that are found in a specific habitat (the area where a species is found). Organisms, populations, and communities have a range of tolerances for each of the limiting factors, a concept known as the law of tolerances. Some organisms have a wide range of tolerances for a limiting factor, such as the concentration of dissolved oxygen in a stream, while others have a narrow range for that same factor.

Biotic limiting factors are perhaps best exemplified by keystone species, species whose activities determine the structure of the community. Scientists are just beginning to understand the various ways in which members of the biota affect ecosystem structure.

 Key Terms

atom	autotroph
biome	carnivore
chemosynthesis	chemotroph
community	compound
decomposer	detritus feeder
detritivore	ecosystem
element	energy
entropy	eutrophication
habitat	herbivore
heterotroph	indicator species
keystone species	limiting factor
macroconsumer	macronutrient
matter	microconsumer
micronutrient	molecule
omnivore	organic compound
photosynthesis	phototroph
phytoplankton	population
predator	prey
primary consumer	producer
range of tolerances	scavenger
secondary consumer	species
tertiary consumer	

Environmental Success Story

Businesses are becoming more environmentally aware and finding out that being "green" can also rake in the green. For example, UPS announced in November 1998 an action plan which cuts across the spectrum of its express packaging. Compared to the previous packaging technique, the improvements — which include reusable envelopes — reduce air pollution by 50 percent, wastewater discharge by more than 15 percent, and energy use by 12 percent. The changes will save UPS more than $1 million annually. In addition, UPS has eliminated the use of bleached paper in all express packaging, has nearly doubled the amount of post-consumer recycled material in the UPS box, and is using at least 80 percent post-consumer recycled material in the Express Letter.

Source: "United Parcel Service — Alliance for Environmental Innovation Overnight Shipping Packaging Project: Project Summary." *The Alliance for Environmental Innovation* (November 1998). Available: http://www.edfpewalliance.org/ups_project_ summary.htm. 18 February 1999.

True/False

1. A community is an area characterized by a dominant vegetation type. T F

2. An ecosystem is self-sustaining because it is a completely closed system. T F

3. Abiotic components of an ecosystem include chemotrophs, detritivores, and decomposers. T F

4. Heterotrophs obtain energy from autotrophs. T F

5. The Earth is an open system because it receives energy from the sun. T F

Fill In the Blank

1. The physical environment in which an organism lives is its _____.

2. Ecosystems are composed of _____ and _____ components.

3. Molecules made up of two or more elements are known as _____.

4. Chemicals needed by organisms in relatively large quantities for the construction of proteins, fats, and carbohydrates are called _____.

5. Organisms such as grasshoppers and deer, which eat only plant matter, are known as _____ or _____.

6. Earthworms and shrimp, which live off the decaying fragments of other organisms, are examples of _____ or _____.

Multiple Choice

Choose the best answer.

1. The region of the eastern U.S. characterized by temperate deciduous forest is an example of a(n)

    A. ecosystem.

    B. biome.

    C. community.

    D. abiotic component.

2. Of the total amount of the sun's energy that reaches the Earth each day, _____ is captured through photosynthesis.

    A. 5-10 percent

    B. 2-5 percent

    C. 1-2 percent

    D. less than 1 percent

3. Decomposers play a major role in an ecosystem by

    A. reducing complex organic matter to inorganic matter.

    B. returning nutrients to a form which can be used by producers.

    C. contributing to the accumulation of nutrients in soil.

    D. All of the above are true.

4. The law which states that during a physical or chemical change energy is neither created nor destroyed is the

    A. first law of energy.

    B. second law of energy.

    C. law of the minimum.

    D. law of tolerances.

5. All of these are examples of a biome except:

    A. tropical savanna.

    B. alpine meadows.

    C. deserts.

    D. temperate grasslands.

Short Answer

1. Describe the levels at which ecology is studied.
2. What is entropy?
3. What are elements?
4. What are the six macronutrients?
5. What is eutrophication?

Thought Questions

Develop a complete answer for each of the following.

1. How do biotic and abiotic components interact to regulate the structure of ecosystems?
2. How do limiting factors affect the structure of an ecosystem?
3. Explain the roles of producers, consumers, and decomposers in an ecosystem.

Related Concepts

Describe the relationship. (There may be more than one.)

BETWEEN...	AND...
heterotrophs	herbivores
chemotrophs	autotrophs
limiting factors	eutrophication
chemical fertilizers	limiting factors
predators	keystone species

Did You Know...?

Campuses with student housing generate an average of 820 pounds of waste per student, per year.

Suggested Activities

1. Observe a terrestrial or aquatic ecosystem first-hand for a few weeks. Identify the autotrophs, primary consumers, secondary consumers, detritus feeders, and microconsumers present in the ecosystem. Describe the interactions between them.

2. Study an aquatic ecosystem (pond, lake, stream, river). How is it being used by humans? How have these uses affected its ecological health?

3. Research the demise of dinosaurs. Which limiting factors are thought to have contributed to their extinction?

Chapter 4
Ecosystem Function

 Chapter Outline

I. **What Is Primary Productivity?**

   A. GPP – R = NPP

   1. Gross primary productivity (GPP) is the total amount of energy fixed by autotrophs over a given period of time.

   2. Respiration (R) is the process by which organisms utilize the chemical energy stored in food.

      a. Aerobic — requires oxygen

      b. Anaerobic — does not require oxygen

   3. Net primary productivity (NPP) is the total amount of energy fixed each year at the producer level minus the amount producers need for their own life processes.

II. **How Does Energy Flow Through A Community?**

   A. Food Chain — Community of Organisms Formed by Trophic Levels

   1. First Trophic Level — Producers

   2. Second Trophic Level — Primary Consumers (Herbivores)

   3. Third Trophic Level — Secondary Consumers (Carnivores)

   4. Fourth Trophic Level — Tertiary Consumers

   5. Decomposers (Microconsumers) operate at each trophic level.

   B. Food Web

   1. Interlocking Food Chains

   2. Defines feeding relationships, movement of energy/materials

III. **How Does Energy Flow Affect the Structure of an Ecosystem?**

   A. Pyramid of Energy

   1. Production, use, and transfer of energy between trophic levels

   2. 10 Percent Rule

      a. Applied to consumption of organisms at one trophic level by organisms at another level

        b. In general, 90 percent of available energy is lost as heat.

        c. Only 10 percent of available energy is transferred at each step.

   B. Pyramid of Biomass

      1. Depicts the total amount of living material at each trophic level

      2. Total biomass tends to become smaller at each level.

      3. Size of each individual tends to become larger.

   C. Pyramid of Numbers

      1. Depicts relative abundance of organisms at each trophic level

      2. Organism at lower levels are found in greater numbers than organisms at higher levels (for example, herbivores are present in greater numbers than carnivores).

IV. How Do Materials Cycle through an Ecosystem?

   A. Biogeochemical Cycling

      1. Cycling involves biological, geologic, and chemical factors.

      2. Three Categories

        a. Hydrologic

        b. Gaseous

        c. Sedimentary

   B. Hydrologic Cycle

      1. Cycling of water from hydrosphere to atmosphere and return

      2. Quality and availability of water are variable.

   C. Gaseous Cycles

      1. Primarily occur in the atmosphere

      2. Most significant gaseous cycles include:

        a. Carbon cycle

        b. Oxygen cycle

        c. Nitrogen cycle

   D. Sedimentary Cycles

      1. Involve materials that move from land to oceans and back

      2. Most significant sedimentary cycles include:

        a. Phosphorous cycle

        b. Sulfur cycle

V. What are Ecosystem Services?

    A. Definition

        1. Functions or processes of a natural ecosystem that provide benefits to human societies

    B. Role and Importance

        1. Products such as fiber, pharmaceuticals, and fuelwood

        2. Services such as the maintenance of biological diversity, seed dispersal, pollination, maintenance of soils, and others

Learning Objectives

After learning the material in Chapter 4, you should be able to:

1. Define gross and net primary productivity.

2. Trace the flow of energy through the biotic community.

3. Describe how energy flow affects the structure of an ecosystem.

4. Explain, in general, how materials cycle in an ecosystem.

5. Explain how the flow of energy and the cycling of matter bind together the structural components of an ecosystem.

Did You Know...?

A dripping faucet can waste up to 2,000 gallons of water a month.

 ## Key Concepts

Read this summary of Chapter 4 and identify the important concepts discussed in the chapter.

The biotic and abiotic components of the biosphere are inseparable, bound together by a complex and delicately balanced web of biological and physical processes that regulate the flow of energy and the cycling of materials. Ecosystems constantly receive energy from the sun, which is the lifeblood that fuels the Earth's biomass.

The release of energy from fuel molecules (carbohydrates and lipids) is called respiration. All organisms respire, but only phototrophs can carry out photosynthesis. Just 0.1 to 0.3 percent of the energy input from the sun is actually captured by phototrophs through photosynthesis, yet that small fraction results in billions of tons of living matter, or biomass.

The gross primary productivity is the total amount of energy fixed by autotrophs over a given period of time. The amount of energy available for storage after the producer's own needs are met (through respiration) is the net primary productivity.

Food chains and food webs represent the feeding relationships and the movement of energy and materials among the organisms of the biotic community. In every ecosystem, some consumers feed solely upon producers, while others feed upon both producers and consumers, and still others feed only upon other consumers. The producers and different kinds of consumers are said to occupy different trophic, or feeding, levels. Producers occupy the first trophic level; primary consumers, or herbivores (plant eaters), occupy the second; secondary consumers, including both omnivores (plant and meat eaters) and carnivores (meat eaters), occupy the third; and tertiary consumers occupy the fourth. A food chain is a simplified illustration of the way in which energy and materials move through an ecosystem. That movement is more accurately depicted by food webs, interlocking chains woven into complex associations.

The flow of energy determines trophic relationships and thus affects the structure of the ecosystem. The pyramid of energy represents the production, use, and transfer of energy from one trophic level to another. Generally, only about 10 percent of the available energy is transferred to the next successive trophic level; the rest is lost to the environment as low-quality heat. The pyramid of biomass represents the total amount of living material at each trophic level, and the pyramid of numbers depicts the relative abundance of organisms at each trophic level. In general, lower levels contain a greater number of organisms than successive levels.

Unlike energy, materials cycle through ecosystems and are used over and over again by the biotic community. The processes by which materials cycle involve living organisms as well as geologic and chemical processes. For this reason, they are known as biogeochemical cycles. The hydrologic cycle accounts for the movement and cycling of water from the hydrosphere and lithosphere to the atmosphere and back to the hydrosphere and lithosphere. Gaseous cycles, which take place primarily in the atmosphere, include the carbon and nitrogen cycles. (Human activities appear to be altering the global nitrogen cycle in adverse ways. The applications of cultural sources of nitrogen include fertilizers, the cultivation of nitrogen-fixing crops, and the combustion of fossil fuels.) Sedimentary cycles, which involve materials that move chiefly from the lithosphere to the hydrosphere and back to the lithosphere, include phosphorous and sulfur.

Energy flow and materials cycling are two examples of ecosystem services — functions or processes of a natural ecosystem that provide benefits to human societies. Others include flood control, crop pollination, and the production of goods such as pharmaceuticals, fuelwood, and fiber. Human activities threaten the continued integrity of ecosystem services. Of greatest concern are the destruction of natural habitats, invasion of nonnative species, loss of native biodiversity, overfishing, and alteration of biogeochemical cycles.

✓ Key Terms

10 percent rule

anaerobic respiration

biomass

detritus food web

estuary

food web

gross primary productivity

net primary productivity

nitrogen fixation

pyramid of biomass

pyramid of numbers

sedimentary cycle

trophic level

aerobic respiration

biogeochemical cycle

carbon cycle

ecosystem services

food chain

gaseous cycle

hydrologic cycle

nitrogen cycle

phosphorus cycle

pyramid of energy

respiration

sulfur cycle

Environmental Success Story

The Central & South West Corporation (CSW) in Tulsa, Oklahoma, has made wind and solar energy a main part of its $17 million renewable energy development program. A public utility holding company, CSW has been successful in showing that these technologies may be used to meet a large portion of our electricity needs. The wind farm has produced over 5.6 million kilowatt-hours (KWH) and various solar photovoltaic systems have produced over 460,000 KWH — all without using fossil fuels! The savings so far roughly equal 15,000 barrels of oil and 3,600 tons of coal. Additionally, over 4,000 tons of carbon monoxide emissions have been avoided.

Source: Renewable Energy Development. "Seventh Annual National Awards for Environmental Sustainability Winners: Renewable Energy." *Renew America* (22 January 1997). Available: http://solstice.crest.org/environment/renew_america/winners.html. 18 February 1999.

True/False

1. Between 0.1 and 0.3 percent of the sun's energy generates the hundreds of billions of tons of biomass on Earth.   T   F

2. Nitrogen is a major component of genetic material, energy   T   F
   materials, and cellular membranes.

3. Human-made sources contribute more sulfur to the atmosphere   T   F
   than natural sources, such as volcanoes.

4. Energy cycles and materials flow through ecosystems.   T   F

5. In general, higher trophic levels represent a greater amount of   T   F
   stored energy, a greater amount of biomass, and a greater
   number of organisms than do lower levels.

Fill in the Blank

1. _____ is the total amount of energy fixed by autotrophs over a given period of time.

2. The process by which energy from fuel (food) is released is _____.

3. _____ is the process by which some kinds of bacteria convert free nitrogen to nitrates or ammonium.

4. The most significant gaseous cycles are the _____, _____, and _____ cycles.

5. An increase in atmospheric CO_2 may cause a climatic change known as _____.

Multiple Choice

Choose the best answer.

1. Of the approximately 1.5 million kcal/m^2/yr which reach the Earth from the sun, about 34 percent

   A. is reflected back to space.

   B. is absorbed by the biosphere.

   C. heats the land and warms the atmosphere.

   D. generates wind currents.

2. Nitrogen is converted to a usable form by

   A. bacteria living in the soil or in plant roots.

   B. lightning.

   C. decomposers during respiration.

   D. All of the above are true.

3. The nutrient that is a major structural component of shells, bones, and teeth is

   A. oxygen.

   B. ammonium.

   C. phosphorus.

   D. nitrogen.

4. The most common element in living tissue is

   A. carbon.

   B. nitrogen.

   C. oxygen.

   D. hydrogen.

Short Answer

1. What is respiration?
2. What is the pyramid of energy?
3. What is the pyramid of numbers?
4. What is the pyramid of biomass?
5. How does the biosphere recycle the carbon dioxide that is produced by processes such as respiration and combustion of fossil fuels?
6. Explain the difference between NPP and GPP.

Thought Questions

Develop a complete answer for each of the following.

1. Which ecosystems are the most productive (efficient at producing energy)? Why? Compare these to less productive ecosystems.
2. Why is "food web" a more accurate term than "food chain"? Why is a detritus food web a particularly good example of this? Why are there usually no more than three or four trophic levels in a given ecosystem?
3. Explain the implications of the 10 percent rule for human diets. How does the typical diet in the U.S. compare with the typical diet in China?
4. Describe the major biogeochemical cycles. What happens when the cycling is interrupted in some way?

▪ Related Concepts

Describe the relationship. (There may be more than one.)

BETWEEN...	AND...
trophic level	food chain
10 percent rule	pyramid of energy
net primary productivity	gross primary productivity

Did You Know...?

Forty percent of all solar energy captured through photosynthesis by terrestrial plants is being used by humans.

 ## Suggested Activities

1. Take note of the food you eat during the next week. At what trophic level do you normally eat? How might you change your eating habits to be more energy efficient?

2. What is your role in biogeochemical cycling? Try to identify all the things you do that move materials through the cycles. For example, if you wash your laundry with a detergent containing phosphates, you are cycling phosphorus. By driving to class, you are contributing carbon dioxide and nitrous oxides (among other things) to the air.

3. Eat less meat. Prepare meals from fruits, vegetables, and grains.

Chapter 5
Ecosystem Development and Dynamic Equilibrium

 Chapter Outline

I. What Causes Ecosystems to Change?

   A. Natural Forces

   B. Human Activity

II. How Do Ecosystems Develop?

   A. Development Through Succession

      1. Process by which an ecosystem matures

      2. Progresses to climax community

         a. Organisms best adapted for conditions in defined area

         b. Production equals respiration — little net production and no further increase in biomass

      3. Occurs in both aquatic and terrestrial environments

   B. Primary Succession

      1. Development of ecosystem in area once devoid of organisms

      2. In temperate region (deciduous forest), the stages consist of:

         a. Lichen pioneer community

         b. Moss community

         c. Herbaceous community

         d. Shrub community

         e. Tree community

         f. Climax forest or equilibrium community

   C. Secondary Succession

      1. Changes that occur after an ecosystem has been disturbed (usually a result of human activity)

      2. In temperate region (deciduous forest), the stages include:

         a. Annual weed community

b. Perennial weed community

  i. Goldenrods as example

  ii. Food source for a variety of insects

  iii. Galls — protect insect eggs

  iv. Support rich diversity of life

c. Shrub or young tree community

d. Young forest community

e. Climax forest or equilibrium community

D. Ecotone

1. Area where different communities meet

2. Stages of succession blend

3. Zone of transition and intense competition

4. Regions supporting great diversity of life

E. Succession in Ponds, Lakes, and Wetlands

1. Nutrient enrichment, and thus eutrophication, may be caused by natural factors or human activity.

2. Eutrophication — maturation of lake or pond due to nutrient enrichment

3. Usually proceeds to terrestrial climax community

4. Marsh may be the climax community (for example, the Everglades)

III. What Is Dynamic Equilibrium?

A. Ecosystems and Communities Characterized by Both Change and Stability

1. Dynamic — constantly undergo changes

2. Equilibrium — condition of balance or stability

B. "Stability" of Ecosystems Based in Part on Feedbacks

1. Positive Feedback — continues a particular trend

2. Negative Feedback — reverses a particular trend

IV. What Factors Contribute to Dynamic Equilibrium?

A. Species Interactions

1. Cooperation

   a. Intraspecific — aids survival of entire group

2. Competition

a. Occurs when two or more individuals vie for resources

b. Intraspecific — between members of the same species

c. Interspecific — between members of different species

   i. Niche

   ii. Competitive exclusion principle

3. Symbiosis

   a. Mutualism — association of two species in which both benefit

   b. Commensalism — association of two species in which one benefits and the other neither benefits nor is harmed

   c. Parasitism — association of two species in which one benefits and the other is harmed

4. Predation

B. Nonhuman Population Dynamics

1. Five Phases Of Population Growth

   a. Lag phase

   b. Slow growth

   c. Log (exponential) growth

   d. Equilibrium

   e. Decline

2. Biotic Potential

   a. Maximum growth rate a population can achieve

   b. Limited by factors known as environmental resistance

3. Two Types Of Environmental Resistance

   a. Density-dependent, or biotic

   b. Density-independent, or abiotic

C. Species Diversity

1. Complex relations between diversity and dynamic equilibrium

   a. Diversity-stability hypothesis

   b. Rivet hypothesis

   c. Redundancy hypothesis

   d. Null hypothesis

2. Illustrates lack of understanding of ecosystem relationships

Learning Objectives

After learning the material in Chapter 5, you should be able to:

1. Explain how ecosystems develop over time through the process of succession.
2. Distinguish between primary and secondary succession.
3. Describe what ecologists mean by dynamic equilibrium.
4. Explain how feedbacks, species interactions, and population dynamics influence dynamic equilibrium.

Did You Know...?

Biologists estimate that as many as five million small mammals die each year in discarded beer bottles.

 ## Key Concepts

Read this summary of Chapter 5 and identify the important concepts discussed in the chapter.

Nature is not static; all ecosystems change and all undergo ecological succession, the gradual, sequential, and somewhat predictable change in the composition of the biotic community. This maturation process is also known as ecosystem development, a term that accounts for the accompanying modifications in the physical environment. Primary succession is the development of a community in an area previously devoid of organisms; opportunistic species known as pioneer organisms invade the area and colonize it. Secondary succession is the change in community types after an area has been disturbed, generally by human activity. Successional stages are not discrete; they tend to blend into one another. For any defined geographic area, as succession proceeds, the species composition of the community changes until the association of organisms best suited to the physical conditions of that area is reached. This association of organisms is known as the climax community.

Ecosystems continually react to change and disturbance, thereby maintaining a dynamic equilibrium, or dynamic steady state. Ecosystems maintain a dynamic equilibrium either by resisting change (inertia) or by restoring structure and function after a disturbance (resilience). Three factors that enable the ecosystem to maintain dynamic equilibrium are feedback, species interactions, and population dynamics. Positive feedback continues a particular trend and can be disruptive, while negative feedback reverses a trend and therefore tends to stabilize the system. Species interactions include predator-prey relationships, competition, and cooperation. Competition and cooperation can take place between individuals of the same species or between individuals of different species. When two species occupy similar ecological niches — defined as an organism's functional role within the community — interspecific competition becomes particularly keen. Situations in which interspecific competition leads to the exclusion or (local) extinction of one of the species illustrate the principle of competitive exclusion. Interspecific cooperation includes symbiosis, the intimate association of two dissimilar species regardless of the benefits or lack of benefits to both species. Mutualism is a symbiotic relationship in which both species benefit. An association in which one species benefits and the other neither benefits nor is harmed is an example of commensalism. Parasitism occurs when one species benefits and the other is harmed.

Population dynamics also contribute to ecosystem stability. In laboratory studies, scientists have found that populations of organisms commonly exhibit five phases of growth: lag; slow growth; log, or exponential, growth; equilibrium; and decline. In contrast, natural populations fluctuate over time around the carrying capacity of the environment, the number of individuals the environment or habitat can best support. Limiting factors, collectively known as environmental resistance, prevent a population from realizing its biotic potential, that is, the maximum growth rate the population could achieve given unlimited resources and ideal environmental conditions. Environmental resistance factors may be density-dependent (effect is greater when the population density is high) or density-independent (population density does not play a role). Density-independent factors set upper limits on populations.

Whether or not biological diversity contributes to ecosystem stability is a question studied and debated by ecologists. The diversity-stability hypothesis and the rivet hypothesis both suggest that biodiversity enhances an ecosystem's resistance to disturbance, while the redundancy hypothesis suggests that only minimum diversity is needed for an ecosystem to function properly. The null hypothesis maintains that there is no relationship between diversity and ecosystem stability.

Key Terms

biotic potential	carrying capacity
climax community	commensalism
competitive exclusion principle	density-dependent factor
density-independent factor	diversity-stability hypothesis
dynamic equilibrium	ecological succession
ecosystem development	ecotone
environmental resistance	inertia
interspecific competition	intraspecific cooperation

mutualism

niche

parasitism

positive feedback

redundancy hypothesis

rivet hypothesis

symbiosis

negative feedback

null hypothesis

pioneer species

primary succession

resiliency

secondary succession

Environmental Success Story

The history of pollution control shows that the benefits usually exceed the costs. In 1972, only about one-third of American lakes and rivers were safe for fishing and swimming; today almost two-thirds are safe, a stunning improvement that occurred even during sustained economic growth. Since 1970, smog in the U.S. has diminished by nearly one-third, even though the number of cars has almost doubled. Most studies suggest the health benefits of declining smog exceed the price of controls by a clear margin — as much as $40 in health care costs saved for every $1 in anti-smog investment, according to a recent EPA analysis.

Source: Easterbrook, Gregg. "Greenhouse Common Sense: Why Global Warming Economics Matters More Than Science." *U.S. News Online.* Available: http://www.usnews.com/utils/search. 2 February 1999.

True/False

1. The first community to occur in primary succession is composed of pioneer species. T F

2. Ecotones have greater diversity than the adjacent habitats. T F

3. The climax community of aquatic ecosystems is *always* a marsh or swamp community. T F

4. In general, negative feedback is destabilizing to an ecosystem. T F

5. Density-independent factors tend to set upper limits on population size. T F

Fill in the Blank

1. The process by which an ecosystem matures is _____.

2. The ability of a natural system to resist change is known as _____.

3. A(n) _____ is an organism that lives in the body cavity, organs or blood of another organism, causing it harm.

4. The first phase of population growth in a controlled environment, in which the population does not grow, is called the _____.

5. An environment's _____ is the population size it can best sustain over time.

Multiple Choice

Choose the best answer.

1. The zone of transition and intense competition where different communities meet is a(n)

    A. ecotone.

    B. pioneer community.

    C. biome.

    D. niche.

2. Two species competing for the same habitat in a community is an example of a(n)

    A. specialist niche.

    B. intraspecific competition.

    C. interspecific competition.

    D. competitive exclusion principle.

3. In controlled lab environments, animal populations pass through five phases. The first three phases are (in order)

    A. lag, exponential growth, slow growth.

    B. lag, dynamic equilibrium, exponential growth.

    C. slow growth, exponential growth, dynamic equilibrium.

    D. lag, slow growth, exponential growth.

4. An example of commensalism is

    A. remora fish that attach to sharks.

    B. liver flukes in cattle.

    C. ticks on sheep.

    D. lichens.

5. Mutualism is when two species interact and

   A. one is harmed and the other benefited.

   B. both are either harmed or benefited.

   C. one is benefited, the other is neither harmed nor benefited.

   D. both are benefited.

Short Answer

1. What is stress?

2. What is a climax community?

3. What is the difference between primary and secondary succession?

4. What is the role of pioneer organisms in ecosystem development?

5. What is the competitive exclusion principle?

Thought Questions

Develop a complete answer for each of the following.

1. Describe primary succession from bare rock to climax forest in a temperate zone. What determines the dominant species in the climax community? What stops a climax community from further succession?

2. Explain the concept of dynamic equilibrium. In what ways does it pertain to the phrase "balance of nature"?

3. How do feedback, species interactions, and population dynamics contribute to the dynamic equilibrium of ecosystems?

4. Why is interspecific competition most intense when two species occupy similar ecological niches?

5. Why do natural populations never reach their biotic potential?

▪ Related Concepts

Describe the relationship. (There may be more than one.)

BETWEEN...	AND...
niche	habitat
diversity	ecosystem stability
specialist niche	generalist niche
density-dependent factor	density-independent factor
pioneer species	climax community

Did You Know...?

Asphalt made with old tires has a life span three times longer than concrete.

 ## Suggested Activities

1. Observe an old field or other disturbed area for evidence of succession. What species do you find there? What stage of succession is occurring?

2. Locate an ecotone. Which species are competing with one another? Can you identify any transitional species?

3. Try to buy products that are less toxic to the environment when produced. For example, use vinegar and water as a replacement for glass cleaner.

Chapter 6
Ecosystem Degradation

Chapter Outline

I. **What Is Ecosystem Damage?**

   A. Adverse Alteration of System's Integrity, Diversity, Productivity

   B. Major Cause: Pollutants

   C. Four Factors Determine the Extent of Damage

   1. Effect of Pollutant

      a. Acute — effects occur immediately, are easily detectable

      b. Chronic — long-term effects, unnoticed for years

      c. Bioaccumulation — storage of chemicals in an organism in higher concentrations than are normally found in the environment

      d. Biomagnification — accumulation of chemicals in organisms in increasingly higher concentrations at successive trophic levels

   2. How Pollutant Enters the Environment

      a. Point-source — identifiable, specific source

      b. Nonpoint-source — from wide area, hard to control

      c. Cross-media — moves from one medium (air, water, soil) to next

   3. Quantity of Pollutant

   4. Persistence of Pollutant: Does it build up in natural systems over time?

      a. Biodegradable — substance that can be broken down and made harmless by organisms in the ecosystem

      b. Nonbiodegradable — substance that cannot be broken down by organisms in the ecosystem

II. **What Is Ecosystem Disruption?**

   A. Rapid Change in Species Composition Traced Directly to Human Acts

   B. Major Causes Include:

   1. Persistent Pesticides

   2. Introduction of New (Exotic) Species

   3. Construction of Dams, Roadways, Pipelines, and Other Structures

        4. Overexploitation of Resources

III. **What Is Ecosystem Destruction?**

    A. Replacement of a Natural System by a Human System

    B. Major Causes Include:

        1. Urbanization

        2. Transportation

        3. Agriculture

IV. **What Is Desertification?**

    A. Land Degradation in Arid, Semi-Arid, and Dry Subhumid Regions

        1. Expands or creates desert conditions in areas where those conditions do not normally occur

    B. Results Mainly from Adverse Human Impact

        1. Overgrazing

        2. Overcultivation

        3. Deforestation

        4. Poor Irrigation Practices

V. **What Is Deforestation?**

    A. Cutting Down and Clearing Away of Forests

        1. Results in erosion and loss of native forest species

    B. Fragmentation

        1. Frontier Forests — expansive tracts of contiguous forest largely untouched by human activities

        2. Fragmented Forests — patchwork of cropland, logging roads, and smaller, discrete forest areas

            a. Do not appear to function as intact ecosystems do

            b. Widespread suffering of resident species

    C. Tropical Rain Forests

        1. Home to 3-30 million species of plants and animals

        2. Moderate temperatures of entire planet

        3. Loss related to political, social, and economic factors

    D. Coastal Temperate Rain Forests of the United States and Canada

        1. Home to very old, very large trees

2. Increased pressure to harvest from logging community

Learning Objectives

After learning the material in Chapter 6, you should be able to:

1. List and define the five Ds of ecosystem degradation.
2. Describe what is meant by ecosystem damage and define pollutant.
3. Explain how human activities cause disruption of natural systems.
4. Identify the primary cause of ecosystem destruction.
5. Define desertification and identify three human activities that contribute to it.
6. Define deforestation and identify its primary causes.

> **Did You Know...?**
>
> Eighty percent of the forests that originally covered the Earth have been cleared, fragmented, or otherwise degraded.

 ## Key Concepts

Read this summary of Chapter 6 and identify the important concepts discussed in the chapter.

Ecosystems become degraded when human activities alter environmental conditions in such a way that they exceed the range of tolerances for one or more organisms in the biotic community. A degraded ecosystem loses some capacity to support the diversity of life forms that are best suited to its particular physical environment. The five Ds of ecosystem degradation are damage, disruption, destruction, desertification, and deforestation.

Damage occurs when the integrity of natural systems is altered. A pollutant is a substance or form of energy, such as heat, that adversely alters the physical, chemical, or biological quality of natural systems or that accumulates in living organisms in amounts that threaten their health or survival. Four factors determine the damage done by a pollutant: the effect of the pollutant, how it enters the environment, the quantity of the pollutant, and its persistence.

A rapid change in the species composition of a community that can be traced directly to a specific human activity is known as a disruption. Activities that may lead to a change in species composition include the use of persistent pesticides, the introduction of new species into an area (usually accidental), the construction of dams or highways, and the overexploitation of resources.

A more common form of natural system degradation is ecosystem destruction. A natural system is destroyed when it is replaced by a human system. Perhaps the most serious consequences of natural system destruction are habitat loss and the loss of the functions and services provided by natural systems — functions that human systems cannot duplicate.

Two specific types of natural system degradation are desertification — land degradation in arid, semi-arid, and dry subhumid regions resulting mainly from adverse human impact — and deforestation — the cutting down or clearing away of forests. Fragmentation results in further ecosystem degradation by segmenting an intact frontier forest into smaller, less functional patches. Overgrazing, overcultivation, deforestation, and poor irrigation practices are the major causes of desertification. The major causes of deforestation are conversion to farmland and pastureland; the demand for fuel, timber, and paper products; and the construction of roadways. In some areas of the world, desertification and deforestation are occurring at alarming rates.

 ## Key Terms

acute pollution effects	anthropogenic
bioaccumulation	biodegradable
biomagnification	chronic pollution effects
cross-media pollutant	deforestation desertification
ecosystem damage	ecosystem destruction
ecosystem disruption	frontier forests
fragmentation	nonbiodegradable
nonpoint source	persistent pollutant
phototoxicity	point source
pollutant	salinization
siltation	synergistic effect

Environmental Success Story

Earth Island Institute in San Francisco, California, has promoted making paper from an alternative source to trees. Kenaf, a plant native to Africa and related to hibiscus and okra, generates more pulp per acre than do forests or tree farms. In 1996, the Earth Island Journal successfully published the first U.S. magazine printed on 100-percent plant paper, and the group is now promoting kenaf in the media as an "environmentally preferred paper product."

Source: Earth Island Institute. "Seventh Annual National Awards for Environmental Sustainability Winners: ReThink Paper." *Renew America* (22 January 1997). Available: http://solstice.crest.org/environment/renew_america/winners.html. 18 February 1999.

True/False

1. Persistent pollutants can be both biodegradable and nonbiodegradable.   T   F
2. The primary cause of species extinction is habitat loss.   T   F
3. Though expensive, the rehabilitation of ecosystems that have undergone desertification is usually possible.   T   F
4. Worldwide, the most common cause of deforestation is the demand for timber and paper products.   T   F
5. DDT is an example of a persistent pollutant.   T   F

Fill in the Blank

1. The Five Ds of natural system degradation are _damage_, _disruption_, _destruction_, _desertification_, and _deforestation_.
2. A(n) _disruption_ is a rapid change in the species composition of a community that can be traced directly to a specific human activity.
3. _destruction_ occurs when a living system is replaced by a human system.
4. A smokestack that releases pollutants into the environment is a(n) _point_ source of pollution.
5. A pollutant which can move from one type of environment (such as air) to another (such as water) is a _cross media pollutant_.

Multiple Choice

Choose the best answer.

1. All of the following are examples of cross-media pollution except:

    A. landfill leachate.

    B. photochemical smog.

    C. acid precipitation.

    D. groundwater contamination.

Student Learning Guide Chapter 6

2. All of the following are examples of destruction except:

   A. conversion of grassland to cropland.

   B. draining of coastal wetlands for agricultural use.

   C. conversion of cropland to housing developments.

   D. All of the above are true.

3. Substances that enter a system in a form unusable by organisms in that system are said to be

   A. nonbiodegradable.

   B. biodegradable.

   C. persistent.

   D. acute.

4. When a chemical is stored in an organism in a higher concentration than is normally found in the environment, it is an example of

   A. biomagnification.

   B. bioaccumulation.

   C. persistent pollutants.

   D. synergism.

5. Worldwide, the primary cause of deforestation is

   A. logging and the paper products industry.

   B. urban sprawl.

   C. construction of roadways.

   D. conversion to agricultural land.

Short Answer

1. What four factors determine how a pollutant affects the environment?
2. What is synergism? Give an example.
3. Which ecosystems are most susceptible to desertification and why?
4. What are the major causes of deforestation?
5. What factors must preventative desertification measures have to be effective?

Thought Questions

Develop a complete answer for each of the following.

1. Describe the five types of natural system degradation and give examples of each. Discuss how their major causes are interrelated.
2. Explain why nonpoint-source pollution is difficult to measure or control.
3. Differentiate between bioaccumulation and biomagnification. How does the concept of trophic levels relate to biomagnification?
4. Discuss the damage caused by the use of persistent pesticides and explain how resistant populations of insects evolve.
5. What are the benefits of forests and what are the threats of deforestation? Why are tropical and coastal temperate rain forests of particular concern?

▢▨■ Related Concepts

Describe the relationship. (There may be more than one.)

BETWEEN...	AND...
irrigation	desertification
persistent pollutant	biomagnification
acute effect	chronic effect
ocean dumping	oxygen depletion
persistent pesticide populations	resistant insect

Did You Know...?

The world is losing 70 acres of forests every minute.

Suggested Activities

1. Identify one or more non-native species in your area. How were they introduced? What impact have they had (or, are they having) on the ecosystem?
2. Choose an area of the world that is threatened by desertification or deforestation. Find out more about the causes of the damage, and outline an environmentally sound management plan for the ecosystem, including strategies for eliminating the causes.
3. Track the history of an endangered or extinct species. What kinds of ecosystem damage caused the species' decline?

Student Learning Guide Chapter 6 55

Chapter 7
Applying Ecological Principles

 Chapter Outline

I. **What Is Applied Ecology?**

   A. Definition

   1. Scientific discipline that measures and attempts to predict the ecological consequences of human activities and to recommend ways to limit damage to and restore ecosystems

   B. Five Subdisciplines

   1. Disturbance Ecology

      a. Predicts impact of stress on natural systems

      b. Aids in the formation of environmental impact statements (EIS)

   2. Restoration Ecology

      a. Concerned with repairing damage to ecosystems caused by human activities

      b. Maintains present diversity of species and ecosystems

      c. Increases knowledge base to restore and manage other systems

      d. Restoration techniques

         i. Artificial reefs

         ii. Prescribed burns

   3. Landscape Ecology

      a. Earth as patchwork of individual ecosystems

      b. Study of ecosystem distribution, function, and change

      c. Emphasizes the relationship between humans and living space

   4. Agroecology (Agricultural Ecology)

      a. Study of agroecosystems with long-term management goal

      b. Emphasis on protecting the health of the soil

   5. Ecotoxicology (Ecological Toxicology)

      a. Study of the effect of toxins on ecosystems

b. Attempts to understand, predict, and monitor the effects of toxins in order to alleviate the impact of pollution

II. What Is Conservation Biology?

A. Scientific Discipline Dedicated to Protecting, Maintaining, and Restoring the Earth's Biological Diversity

1. Requires knowledge of the complex relationships among the biota
2. Employs a wide range of professionals
3. Goal is action — timely, real-world solutions

III. How Can Computer Models Aid Environmental Management?

A. Predict the Effect of Management Strategies on Ecosystems

B. Understand How Ecosystems Maintain a Dynamic Steady State

C. Help Monitor Changes in Ecosystems Over Time

D. Predict Ecosystem Response to Stress

Learning Objectives

After learning the material in Chapter 7, you should be able to:

1. Define applied ecology and explain how it differs from ecology.
2. Identify five subdisciplines of applied ecology and give examples to illustrate each.
3. Define conservation biology and explain why it takes an interdisciplinary approach to environmental management.
4. Explain how applied scientists use computer models in their work. List the advantages and disadvantages of computer modeling.

Did You Know...?

The Trust for Public Land estimates that four square miles of open space are paved over every day.

 ## Key Concepts

Read this summary of Chapter 7 and identify the important concepts discussed in the chapter.

Five subdisciplines of applied ecology are disturbance ecology, restoration ecology, landscape ecology, agroecology, and ecotoxicology. Disturbance ecology is primarily concerned with assessing the impact of specific stresses on particular organisms, populations, and ecosystems; understanding the causes and effects of change in natural systems; and planning and designing appropriate projects to mitigate the adverse effects of human activities.

Although closely related to disturbance ecology, restoration ecology is a distinct subdiscipline that is driven largely by the values of society. Its three main goals are (1) to repair biotic communities on their original sites or to establish them on other sites if the original sites can no longer be used; (2) to maintain the present diversity of species and ecosystems; and (3) to increase knowledge of biotic communities and restoration techniques. The construction of reefs and the use of fire are important management techniques for restoration ecologists.

Landscape ecology attempts to understand how ecosystems interact to form a larger area. It is holistic in its focus on the connections among various subunits, or patches, of the landscape. Landscape ecology is especially interested in the relations between human society and its living space, both the natural and managed components of the landscape.

Agroecology is concerned with developing methods of food production that are ecologically sustainable. Agroecology often incorporates ideas and practices that reflect a sensitivity to social and environmental concerns. Its roots are in the agricultural sciences, ecology, the environmental movement, studies of the agricultural practices and knowledge of indigenous peoples, and studies of rural development.

Ecotoxicology is a merger of ecology and toxicology. Ecotoxicologists study, monitor, and predict the consequences of a wide variety of pollutants in order to offer suggestions for mitigating pollution effects.

Conservation biology aims to protect, maintain, and restore the Earth's biological diversity. Conserving biological diversity depends on sustaining the complex relationships between all species; therefore, conservation biology emphasizes the interaction of individual species, ecological processes, and human societies. Because conservation biology takes a comprehensive approach to diversity issues, it employs a wide range of professionals. In addition, conservation biology asks all people to become involved in promoting biological diversity by becoming aware of humankind's place in the natural world.

Applied ecology and conservation biology share some common techniques and tools. One of the most important of these is computer models, sophisticated mathematical equations that enable applied scientists to understand how ecosystems respond to stress and to predict the effects of management strategies. Because of the complexity of natural systems, it takes years of careful observation and data collection to develop an accurate computer model. Models have been developed to simulate both terrestrial and aquatic ecosystems and to predict the likely effects of climate change.

 **Key Terms**

agroecology

applied ecology

conservation

disturbance ecology

landscape ecology

preservation

toxicology

agroecosystem

computer model

conservation ecology

ecotoxicology

prescribed burning

restoration ecology

toxin

Environmental Success Story

In 1997, an outbreak of the toxic algae *Pfiesteria piscicida* occurred in the Chesapeake Bay. *Pfiesteria,* also called the "cell form hell," soon became the suspected cause of many distressing symptoms — including nausea, fatigue, and memory loss — reported by dozens of dock and water workers. The *Pfiesteria* outbreak is just one of several ecological troubles brewing in the Bay; others include annual algae blooms, sparse underwater grasses, and low levels of dissolved oxygen. These problems are the likely result of increased industrial pollutants and excess nutrients pouring into the Bay from an expanding coastal human population.

However, there is hope. Scientists say a good candidate to help clean up and rejuvenate the Bay is the renowned — and very much depleted — eastern oyster. Oysters are consummate filter feeders, vacuuming up the algae and suspended silt that cause many of the Chesapeake Bay's problems. In the past decade, Virginia and Maryland, which have jurisdiction over the Bay, have taken steps to restore its oyster population. The two states are dumping more than three million bushels of oyster shells each year in dozens of historic reef areas in an effort to resurrect some old oyster breeding grounds. Computer models predict that an increase in the oyster population of as little as 10 percent of its historic high would improve water clarity and spur the growth of sea grass meadows.

Source: Zimmermann, Tim. "Filter it with Billions and Billions of Oysters." *U.S. News Online* (29 December 1997). Available: http://www.usnews.com/usnews/issue/971229/29oyst.htm. 2 February 1999.

True/False

1. One of the goals of restoration ecology is to allow natural succession to occur.   T   F

2. The emphasis of agroecology is on protecting the health of the soil.   T   F

3. Conservation is the planned management of an area or resource in order to provide for its continued use.   T   F

4. Applied ecology is not concerned mainly with human activities.   T   F

5. Most toxicological work done to date has focused on the effects of specific substances on organisms rather than on communities or ecosystems.   T   F

Fill in the Blank

1. The subdiscipline of applied ecology which assesses the impact of stresses on organisms, populations, and ecosystems is _____.

2. _____ is the subdiscipline of applied ecology that focuses on the relations between human society and its living space.

3. Certain species, called _____, are more sensitive to pollutants than others and are useful in determining when pollutant levels are unsafe.

4. The study of purely ecological phenomena within the crop field is called _____.

5. The wise use and planned management of an area or resource in order to ensure its continued use is known as _____, while _____ is the management of an area in ways that restrict its use to nonconsumptive activities.

Multiple Choice

Choose the best answer.

1. The building of artificial reefs is an example of a technique used in
    A. conservation biology.
    B. restoration ecology.
    C. landscape ecology.
    D. All of the above are true.

2. The ecologist most likely to be involved in developing an environmental impact statement is a(n)
    A. restoration ecologist.
    B. agroecologist.
    C. disturbance ecologist.
    D. preservation ecologist.

3. The primary focus of agroecology is

   A. preserving biological diversity.

   B. protecting the health of the soil.

   C. using integrated pest management.

   D. maximizing the genetic potential of hybrids.

4. Conservation biology focuses on:

   A. researching interactions among the biota.

   B. developing plans of action.

   C. increasing humans' awareness of their place in the natural world.

   D. All of the above are true.

5. Which of the following make the development of computer models a difficult and uncertain process?

   A. lack of necessary data to compare stressed and unstressed systems

   B. disagreement over which problems should be addressed

   C. lack of computers powerful enough to process pertinent data

   D. too little interest from scientists in developing or using computer models

Short Answer

1. List the five subdisciplines of applied ecology.
2. How does applied ecology differ from ecology?
3. What are the three major goals of restoration ecology?
4. What is conservation biology?
5. How do agroecosystems differ from natural ecosystems?

Thought Questions

Develop a complete answer for each of the following.

1. Discuss the use of prescribed burning and other techniques to keep ecosystems in certain successional stages. How do societal values play a role in determining which ecosystems are maintained in this way?

2. How is conservation biology similar to restoration ecology? How is it different?

3. Discuss the use of computer models in applied ecology. What are the benefits of using these models? What are their limitations?

4. Which subdiscipline of applied ecology do you think is most valuable and why?

5. How do social and cultural factors influence conservation biology and each of the five subdisciplines of applied ecology? Consider each area separately and illustrate your answer with examples.

■ Related Concepts

Describe the relationship. (There may be more than one.)

BETWEEN...	AND...
conservation	preservation
indicator species	bioaccumulation
jack pines in Michigan	meadows in Yosemite National Park
landscape ecology	satellite imagery
computer models	scientific hypothesis

Did You Know...?

If everyone in the U.S. recycled the Sunday newspaper, it would save 500,000 trees every week.

 ## Suggested Activities

1. Review an environmental impact statement or study a project for which an EIS is being prepared (check newspapers or periodicals to find a project that interests you). What contributions do you think conservation biology and the five subdisciplines of applied ecology might have made (or might make) to this document?

2. Visit a restored ecosystem in your area. What is its history? What techniques were used to restore it?

UNIT III

An Environmental Imperative:
Balancing Population, Food, and Energy

Chapter 8
Human Population Dynamics

 Chapter Outline

I. **Describing the Human Resource: Biological Characteristics**

   A. Can the Human Population Be Considered A Resource?

   B. The Great Population Debate

   1. "No Population Problem"

      a. Cornucopians: People are the world's ultimate resource.

      b. Marxists: Poverty is the result of distribution problems, not overpopulation.

   2. "There Is a Population Problem"

      a. Malthusians: Population growth, which is exponential, is limited by growth in the food supply, which is arithmetic.

      b. Neomalthusians: In addition to food, other factors (such as shortage of water and space) impose limits on continued growth.

      c. Zero Population Growth: A halt in population growth is needed.

   3. "The 'Population Problem' Is a Complex Issue"

      a. Problems are the result of unequal distribution of resources and high growth rates.

      b. Overconsumption by slow-growth countries is also problematic.

   C. What Is Demography?

   D. Why Do We Study Demography?

   E. How Are Populations Measured?

   1. Number of People

      a. Absolute numbers (world) = total live births − total deaths

      b. Absolute numbers (country or region) = (births + immigration) − (deaths + emigration)

   2. Growth Rates

      a. Crude birth rate (CBR) — number of live births per 1,000 people

      b. Crude death rate (CDR) — number of deaths per 1,000 people

c. Zero population growth (ZPG) — births = deaths; zero growth rate

3. Doubling Times

a. Determined by rule of 70 — 70/current annual growth rate = doubling time in years

b. LDCs have the shortest doubling times.

F. What Factors Affect Growth Rates?

1. Fertility

a. General fertility rate — number of live births per 1,000 women of childbearing age per year (ages 15-49)

b. Age-specific fertility rate — number of live births per 1,000 women of a specific age group per year

c. Total fertility rate — average number of children a woman will bear throughout her life, based on the current age-specific fertility rate and assuming the current birth rate will remain constant throughout her life

d. Replacement fertility rate — fertility rate needed to ensure that each set of parents is "replaced" by their offspring

2. Age Distribution

a. Graphically represented by population profile

i. LDCs have pyramidal-shaped profile; MDCs have more rectangular or columnar profile.

b. Important indicator of future growth rates

i. Population momentum — occurs when large numbers of young people are present in the population, ensuring continued growth even after fertility rates drop

c. Determines a nation's dependency load — the proportion of the population below 15 or above 65 years of age

3. Migration

a. Actual rate of increase

G. How Does Population Growth Affect Economic Development?

1. Every one percent increase in population needs a three percent increase in GNP.

2. High growth rates of LDCs have overwhelmed governments.

II. Describing the Human Resource: Physical Characteristics

A. What Can We Learn from Studying Nonhuman Population Dynamics?

1. Biological Carrying Capacity — maximum number of a particular species that the environment can support

2. Optimum Population Size — number of individuals that the environment can *best* support

B. Carrying Capacity as Applied to Human Populations

1. No standard carrying capacity equation for humans

2. Humans are classified geopolitically, classifications which are ecologically meaningless.

3. Humans can raise carrying capacity of environment through technology.

4. Quality of life (standard of living) separates the calculation of carrying capacity between humans and nonhumans.

   a. Cultural carrying capacity

5. When standards of living drop dramatically, people may become environmental refugees.

III. Describing the Human Resource: Social Characteristics

A. What Do Demographic Statistics Tell About Quality of Life?

1. Population Density

2. Urbanization

3. Life Expectancy (at birth)

4. Infant and Childhood Mortality

   a. Infant mortality rate (IMR) —number of infants under one year of age who die per 1,000 births each year

   b. Childhood mortality rate (CMR): number of children between one and five years of age who die per 1,000 births each year

   c. IMR is best single indicator of a society's quality of life.

   d. Factors that contribute to high IMR and CMR include diarrhea (often caused by disease-infested water), improper weaning, famine, malnutrition, poor health of the mother, and inadequate prenatal care.

IV. What Is the Relationship Between Women's Status and Population Growth?

A. If Women's Status Is Low:

1. Motherhood Only Option

2. Birth Rates Rise

B. Status Determined By:

1. Access to Education
2. Access to Adequate Health Care
   a. Maternal mortality ratio
3. Legal Rights
4. Employment Opportunities Outside Home
5. Wage Earnings
6. Marriage Age
7. Number of Children

C. Low Status of Women Related to Increasing Child Sex Trade
1. Both girls and boys are used as sexual objects, but problem is greater for young girls.
2. Problem is widespread, but it is most severe in Asia and South America.

Learning Objectives

After learning the material in Chapter 8, you should be able to:

1. Explain the basic arguments of each side in the Great Population Debate.
2. Define demography and discuss the demographic statistics pertaining to population size and growth (how populations are measured and what information those statistics reveal).
3. Describe the relationship between carrying capacity and population growth.
4. Discuss how demographic statistics are indicators of the quality of life for a particular nation.
5. Compare population growth and quality of life in the more-developed countries (MDCs) and the less-developed countries (LDCs).

Did You Know...?

In 1997, 80 million people were added to the world's population. Approximately 60 percent were added to Asia alone.

 **Key Concepts**

Read this summary of Chapter 8 and identify the important concepts discussed in the chapter.

Whether or not one defines human population growth as a problem is at the heart of the Great Population Debate. On one side of the debate are, among others, Cornucopians, who believe that human population growth presents no problem and, indeed, that continued growth is desirable because more people mean more consumers, more ideas, and more human inventiveness. On the other side of the debate are Malthusians, proponents of the beliefs credited to Thomas Malthus, an eighteenth-century parson who believed that humans tend to reproduce out of an innate desire to procreate. According to Malthus, that tendency is offset by food production, which is limited; additionally, war, famine, and pestilence act as negative feedbacks that help to limit population growth. The poor are disproportionately affected since they cannot afford to buy food supplies or medicines and are more likely to be sent to war. Neomalthusians agree that there is a link between poverty and family size but argue that food may not be as important a factor as Malthus believed.

The scientific study of the sum of individual population acts is called demography. Demographers study how populations change over time in order to understand the causes and consequences of human population dynamics. A population can be measured in one of three ways: absolute number of people, growth rate, or doubling time. The actual increase (or decrease) in absolute numbers is given as (the number of births + immigration) – (the number of deaths + emigration). The growth rate is defined as the difference between the crude birth rate (number of live births per 1,000 people) and the crude death rate (number of deaths per 1,000 people), expressed as a percentage. When births are equal to deaths, the growth rate is zero, and the condition is known as zero population growth (ZPG). The doubling time is simply the number of years it will take a population to double, assuming the current growth rate remains constant. Most of the people born in the coming decades will live in less-developed countries (LDCs). LDCs all have the fastest doubling times; in contrast, more-developed countries (MDCs) have much longer doubling times (some of them in the thousands of years) because these more industrialized, wealthier nations are growing far more slowly — if they are growing at all. Doubling times are calculated by the rule of 70: 70 divided by the current growth rate is equal to the doubling time.

Growth rates are affected by fertility, age distribution, and migration. Measures of fertility, the actual bearing of offspring, include the general fertility rate, the age-specific fertility rate, the total fertility rate, and the replacement fertility rate.

The age distribution of a population is graphically represented by a population profile, an age structure histogram that employs horizontal bars to depict the number of males and females of each age group. Population momentum occurs when there are large numbers of children living as fertility rates begin to drop; the population, in terms of absolute numbers, continues to grow significantly because there are more people entering their childbearing years. Population profiles for LDCs typically have a pyramidal shape because the largest group within the population is the young (under 15) and the smallest is the elderly (over 65). The age distribution of a population reveals important information about the population's future growth (if there are many young people, the population will continue to experience population momentum) and dependency load (a significant proportion of young or old means a higher dependency load for those who are working). Population growth also affects both GNP and per capita GNP; high growth or a large proportion of dependents (whether young or elderly) places a greater strain on a government to provide basic services (such as roads, schools, and hospitals).

For any country, the actual rate of increase differs from the natural rate of increase because it takes into account migration. The movement of people into a country is immigration; the movement of people from a country is emigration.

Calculating the carrying capacity of the environment for humans (known as the cultural carrying capacity) requires factoring in the quality of life to which a population aspires. Determining cultural carrying capacity is difficult and imprecise but necessary if we are to achieve a balance between the human population and the ability of the environment to sustain life.

Four demographic factors reveal significant information about the quality of life of the majority of people in a given society: population density, how closely people are grouped; urbanization, a rise in the number and size of cities; life expectancy at birth, the average number of years a newborn can be expected to live; and infant and childhood mortality, the number of infants under age one who die each year per 1,000 live births and the number of children between the ages of one and five who die each year per 1,000 live births, respectively. The infant mortality rate (IMR) is widely considered the single best indicator of a society's quality of life. IMRs, whether in LDCs or among the poor in MDCs, could be lowered if all people had access to clean water, sanitation and adequate health care, and if mothers were given the education and health services needed to care for their children.

Improving women's status could help to lower population growth by opening new avenues to females beyond motherhood. Women's status is reflected in access to education, access to adequate health care, legal rights, employment opportunities outside the home, wage earnings, marriage age, and number of children. It is also reflected in the growing child sex trade, in which young girls and (less frequently) young boys are bought, sold, traded, or stolen to be used as sexual commodities. The child sex trade is widespread, but it is most problematic in Asia and South America. Child prostitutes are not a new phenomenon, but their numbers have increased dramatically in recent years. One reason for the boom in the child sex trade is fear of sexually transmitted diseases; brothel customers feel that their chances of contracting Acquired Immune Deficiency Syndrome (AIDS) and other diseases are lower if they are "served" by a young girl or boy.

 **Key Terms**

actual rate of increase	age distribution
age-specific fertility rate	childhood mortality rate
crude birth rate	crude death rate
cultural carrying capacity	demography
dependency load	doubling time
emigration	environmental refugees
fertility	general fertility rate
growth rate	natural rate of increase
immigration	infant mortality rate
life expectancy	maternal mortality rate
migration	natural rate of increase
negative population growth	population density
population momentum	population profile
quality of life	replacement fertility
rule of 70	standard of living
total fertility rate	urbanization
vital statistics	zero population growth

Environmental Success Story

The Population and Habitat Campaign of the National Audubon Society works in the U.S. and abroad to increase funding for international family planning programs. The campaign distributes publications on the cultural and environmental impacts of population growth. It also sponsors Congressional lobby efforts to educate the U.S. government about the link between human population growth and habitat loss. Since 1991, the campaign has pushed Congress to double the U.S. contribution to international family planning.

Source: National Audubon Society. "Eighth Annual National Awards for Environmental Sustainability Winners: Family Planning and Population Stabilization." *Renew America* (22 April 1998). Available: http://solstice.crest.org/sustainable/renew_america/winner98.html. 24 March 1999.

True/False

1. Of the more than five billion people in the world, over four billion live in LDCs.   T   F

2. Migration into a country or region is called emigration.   T   F

3. Africa, western and southern Asia, and Central America are the regions with the highest growth rates.   T   F

4. The carrying capacity for human populations is determined in the same way as nonhuman populations.   T   F

5. The infant mortality rate in the U.S. is much lower than that of other industrialized nations.   T   F

Fill in the Blank

1. The number of live births per 1,000 people is a statistic called the _____.

2. The fertility rate needed to ensure that each set of parents is replaced by their offspring is known as the _____.

3. _____ occurs when there are large numbers of children living as fertility rates begin to drop.

4. To remain economically healthy, a nation's _____ should increase by three percent for every one percent increase in population growth.

5. The _____ rate is the most significant factor that keeps the average life expectancy for LDCs below that of MDCs.

Multiple Choice

Choose the best answer.

1. The world's three most populous countries are

   A. China, Russia, and Indonesia.

   B. China, India, and Russia.

   C. China, India, and the United States.

   D. China, India, and Indonesia.

2. If the crude birth rate is 32/1,000 and the crude death rate is 12/1,000, what is the growth rate?

   A. 4.4 percent

   B. 2.7 percent

   C. 2.0 percent

   D. 0.2 percent

3. World population may reach over _____ by about 2025.

   A. 18 billion

   B. 14 billion

   C. 12 billion

   D. 10 billion

4. The doubling time for a country whose population is growing at 1.4 percent is

   A. 50 years.

   B. 20 years.

   C. 9.8 years.

   D. None of the above is true.

5. A country with zero population growth is

   A. Germany.

   B. Hungary.

   C. Italy.

   D. Denmark.

Short Answer

1. How is the actual population increase in absolute numbers determined for a particular country or region?

2. What is ZPG?

3. What is the doubling time and how is it calculated? Why is it a useful measure?

4. List and briefly explain the three most important measures of fertility.

5. What demographic statistics are used to assess quality of life?

Thought Questions

Develop a complete answer for each of the following.

1. How do Cornucopians, Marxists, Malthusians, Neomalthusians, and advocates of ZPG answer the question, "Is human population growth a problem?" Explain the reasoning behind each of their answers.

2. Explain why "third world" is a value-laden term. How does this label reflect values placed on different human populations? What might be the effects of such labels?

3. What is the growth rate and how is it determined? How do demographers use the growth rate and why is it sometimes misleading?

4. Discuss the factors that contribute to the rapid population growth in LDCs.

5. Contrast the population profile of an LDC with that of an MDC. What social and economic implications do the differences have?

6. What are the factors which distinguish human carrying capacity from nonhuman carrying capacity? Explain the concept of cultural carrying capacity.

7. Discuss the causes and effects of urbanization.

8. How do the low status of women and the rising child sex trade affect population growth? What do they indicate about the value societies and individuals place on human lives and freedoms? How are such human rights issues related to other environmental issues?

■ Related Concepts

Describe the relationship. (There may be more than one.)

BETWEEN...	AND...
natural growth rate	actual growth rate
total fertility rate in MDCs	total fertility rate in LDCs
status of women	population growth
standard of living	resource use
carrying capacity	cultural carrying capacity

 ## Suggested Activities

1. Write a letter to the editor of your local newspaper arguing for your position in the Great Population Debate.

Student Learning Guide Chapter 8

2. What will life be like in the U.S. in 2025, when world population reaches 10 billion? Alternatively, what will life be like in Dhaka, Bangladesh or Mexico City, Mexico? Describe in writing what you imagine.

3. Read *The Ultimate Resource* by Julian Simon.

4. Read *The Population Explosion* by Drs. Paul and Anne Ehrlich.

Did You Know...?

Since 1950, the number of people living in urban areas worldwide has jumped from 750 million to 2.64 billion people.

Chapter 9
Managing Human Population Growth

 Chapter Outline

I. How Has the Human Population Grown Historically?

    A. Early Hunter Gatherers

        1. Nomadic, With a Strong Sense of the Earth

        2. Practiced Intentional Birth Control

    B. Rise of Agriculture

        1. Necessary for Survival

            a. Animals became extinct via predation and altered habitat.

            b. Humans began to cultivate own food.

        2. Agriculture as Impetus for Having Children to Perform Labor

    C. Agriculture Gives Rise to Cities

        1. Food Produced in Country, Consumed in City

            a. Food wastes are no longer returned to soil.

            b. Soil becomes less productive.

        2. Waste of Populations Concentrated in Cities

        3. Population Control in Medieval Societies

            a. Infanticide

            b. Plagues

    D. Industrialization

        1. View of Children During Early Phases of Industrial Growth

            a. Valued as cheap source of income and cheap labor

            b. Exponential growth of populations

        2. By 1900s, Birth Rate in Industrialized World Dropped

            a. Rise in standards of living

            b. Safe and inexpensive means of birth control introduced

            c. Increase in the cost of child rearing

II. What Is the Demographic Transition?

A. Demographic Transition

1. Movement of a nation from high population growth to low population as it develops economically

2. Transition as a Result of Four Stages

   a. Stage 1 — Birth and death rates are both high.

   b. Stage 2 — Death rates fall; birth rates remain high; growth rate rises.

   c. Stage 3 — Birth rates fall as standard of living rises; growth rate falls.

   d. Stage 4 — Growth rate continues to fall to zero or to a negative rate.

B. Demographic Trap

1. Population Path of Most Less Developed Countries (LCDs)

2. "Trapped" in Stage 2 of Demographic Transition

   a. Before 1970, LDCs seemed poised to make transition thanks to economic growth.

   b. Since 1970, economic growth has not kept pace with population.

   c. High birth and low death rates result in explosive population growth.

   d. Downward spiral in standard of living

C. Demographic Fatigue

1. Condition characterized by a lack of financial resources and an inability to deal effectively with threats such as natural catastrophes and disease outbreaks

2. Possibility that countries suffering from demographic fatigue could slip back into stage 1 of demographic transition

III. What Policies Are Used to Control Population Growth?

A. Population Policies

1. Planned course of action or inaction taken by a government

2. Designed to influence choices or decisions on fertility/migration

B. United States Government — Unofficial Pronatalist Policy

1. Income tax deductions for all children

2. Increased benefits for each child born into a welfare family

3. Belief that economy is based on continued population growth

4. Belief that family size should be decided by the family

C. Other Countries — Antinatalist Policies

1. China
2. Peru
3. Philippines

IV. What Are the Arguments in Favor of Controlling Population Growth?

A. Arguments Opposing Any Growth in Population

1. Growth results in worsened living conditions.
2. Environmental degradation (caused by large populations) lowers the carrying capacity.
3. Population growth and pressures are related to violent conflict.

B. Arguments Opposing Rapid Population Growth

1. Negatively impacts economic development
2. Negatively impacts living standards
3. Negatively impacts programs that improve peoples' lives

V. What Is Family Planning?

A. Definition

1. Measures enabling parents to control number of children (if they so desire) and spacing of births

B. Goals of Family Planning

1. Not to limit births
2. For couples to have healthy children
3. For couples to be able to care for their children
4. For couples to have the number of children that they want

C. China's Program

1. Nation With Best Known Population Control Program
2. Reasons Chinese Government Initiated Population Control Measures
    a. Freshwater and food at a premium for nation's population
    b. Country experiencing population momentum
3. Government Perks/Coercive Measures for Citizen Compliance
    a. Free education and health care
    b. Increased personal and family incomes
    c. Increased legal marrying age for women

d. Contraceptives, abortions, and sterilizations free of charge

e. Preferential housing and retirement income

VI. What Methods Are Used To Control Births?

A. Preconception Birth Control Methods

1. Barrier Methods

    a. Condom

    b. Vaginal sponge

    c. Diaphragm

    d. Spermicides

2. Hormonal Contraceptives

    a. Pill

    b. Injections and implants

3. Sterilization

B. Postconception Birth Control Methods

1. Intrauterine Device

2. RU-486 Pill

3. Abortion

C. Contraceptive Use Worldwide

1. People in industrialized countries enjoy easy access to contraceptives while those in LDCs do not.

2. In the U.S., teens and poor woman are least likely to use contraceptives.

3. Severe problems are associated with teen pregnancy.

    a. Teens don't receive the care they need.

    b. Survival rate for babies born to teens is low.

    c. Young age of mother can cause problems with the child.

    d. Teen pregnancy causes greater public expenditures.

Learning Objectives

After learning the material in Chapter 9, you should be able to:

1. Briefly recount how the human population has grown historically.

2. Describe the demographic transition, identify the factors needed for the transition to occur, and explain what is meant by the demographic trap and demographic fatigue.

3. Define population policy and explain how policies can be used to encourage or discourage growth.

4. Summarize the arguments given in favor of controlling or limiting population growth.

5. Discuss the difference between family planning and birth control; give examples of each.

Did You Know...?

If current growth rates remained constant, the world population would double in 50 years and would be about 12 billion by 2050, 24 billion by 2100, and 48 billion by 2150!

Key Concepts

Read this summary of Chapter 9 and identify the important concepts discussed in the chapter.

For most of its tenure on Earth, beginning some 40,000 or 50,000 years ago, *Homo sapiens* existed by gathering wild plants and hunting wild animals. Tribal groups controlled their populations through abstinence from sexual intercourse, birth spacing, and infanticide. Some 10,000 to 12,000 years ago, the last ice age came to an end and the climate began to warm significantly. Human predation and loss of habitat (caused by the warming) led to the extinction of numerous species of large mammals. Human tribes were forced to begin to cultivate more of their foods, and some abandoned their nomadic lifestyle for a more settled lifestyle. Children became valued because they could help with household and agricultural chores. Consequently, with the rise of agriculture, populations began to grow more rapidly, and the environmental impact of human activities increased. Advances in death control ushered in a period of rapid growth. Birth rates in the western world began to fall with the onset of the Industrial Revolution and subsequent rising standard of living, the introduction of safe and reliable means of birth control, and an increase in the cost of child rearing. However, population continued to grow (though more slowly) because there were so many more people in total.

The population path followed by the industrialized nations, the demographic transition, describes the movement of a nation from high growth to low growth. There are four stages in the demographic transition. In stage 1, birth and death rates are both high; birth rates are actually fairly constant but death rates fluctuate as seasonal and cyclic factors (such as disease, harsh or mild weather) come into play. In stage 2, death rates fall but birth rates remain high and the population undergoes rapid growth. In stage 3, economic development improves the society's standard of living and birth rates begin to fall; the growth rate also begins to fall and nears zero. Eventually, the population's growth rate declines to a zero or negative rate, an indicator of a population in stage 4 of the demographic transition.

Nations in the developing world have found it difficult, if not impossible, to follow the path of demographic transition. Much of the growth in the global population in the past fifty years has occurred in LDCs; falling death rates and constant or slightly rising birth and fertility rates are responsible for this growth. Economic development has failed to keep pace with population growth; rapid growth and the environmental deterioration it causes result in a downward spiral in the standard of living. Some developing nations are thus caught in a demographic trap, unable to break out of stage 2 of the demographic transition. Their governments are in a state of demographic fatigue, worn down and financially strapped after decades of struggling to combat the consequences of rapid population growth.

Any planned course of action taken by a government designed to influence its constituents' choices or decisions on fertility or migration can be considered a population policy. A pronatalist policy encourages natality, or births; an antinatalist policy discourages births. Pronatalist or antinatalist policies are developed as a result of the way population changes are perceived.

Family planning is an umbrella term used to describe a wide variety of measures that enable parents to control the number of children they have and the spacing of their children's births. Its goal is not to limit births, though it may be used to do so; rather, the goal of family planning is to enable couples to have healthy children, to care for their children, and to have the number of children they want. China became the first nation to have as an official goal the end of population growth and the subsequent lowering of absolute numbers by a significant amount. From 1969 to 1979, China achieved a transition from high to low birth rates by implementing the strongest family planning measures ever attempted.

Birth control can be achieved through various preconception and postconception methods. Natural forms of contraception, such as rhythm, have a high failure rate relative to many artificial methods. To be as effective as possible, the rhythm method requires a fairly sophisticated knowledge of body processes and cooperation between partners. Other forms of preconception birth control include barrier methods (condoms, spermicides, vaginal sponge, and the diaphragm), hormonal contraceptives (the pill, injections, and implants), and sterilization. Postconception birth control methods include the intrauterine device, the RU-486 pill, and abortion. The abortion controversy revolves around the conflict between religious and moral beliefs about the status of the fetus.

Worldwide, contraceptive use varies widely. People in the industrial nations enjoy easy access to birth control methods, while those in the developing nations, in general, do not. Contraceptive use also varies widely within the United States.

 **Key Terms**

abortion

demographic fatigue

demographic trap

population policy

antinatalist policy

demographic transition

family planning

pronatalist policy

Environmental Success Story

Experts say that less than one-tenth of one percent of the Illinois prairie remains in its original condition. The Fermilab National Accelerator Laboratory is helping to change the tide of this environmental erosion by restoring native tall grass, savannas, woodlands, and wetlands. Since 1975, Fermilab's Ecological Land Management Committee has been improving these native habitats through soil preparation, seeding, selective overseeding, plant surveys, and periodic burning. The more than 1,000 acres involved in this restoration process maintain and protect many native plant and animal habitats, giving rise to an increased number of birds, mammals, reptiles, and insects which were rarely, if ever, seen before the restoration project began. The program also involves local junior high and high school students who participate in plant and soil studies.

Source: Natural Areas Restoration. "Seventh Annual National Awards for Environmental Sustainability Winners: Fermilab." *Renew America* (22 April 1998). Available: http://solstice.crest.org/environment/renew_america/winners.html. 24 March 1999.

True/False

1. The basic goal of family planning is to limit births.   T   F

2. Advances in death control, rather than increased birth rates, led to rapid population growth during the Industrial Revolution.   T   F

3. In general, the nations with the highest average per capita incomes have rates of population growth that are considerably above the world average.   T   F

4. Compared to less educated women, better educated women are more likely to have large families.   T   F

5. Abortion, IUDs, and the RU-486 pill act *after* conception has occurred.   T   F

Fill in the Blank

1. The _____ is the movement of the population of a nation from high growth to low growth.

2. Any planned course of action or inaction taken by a government designed to influence its constituents' choices or decisions on fertility or migration can be considered a _____.

3. _____ includes a wide variety of measures that enable parents to control the number of children they have and the spacing of their children's births.

4. Spermicides, hormonal contraceptives, and sterilization are examples of _____ birth control methods.

5. Nations that are unable to break out of the second stage of the demographic transition are said to be in the _____.

Multiple Choice

Choose the best answer.

1. When people first began to practice agriculture about 10,000 years ago, the population of the Earth was probably about

    A. 1 million.

    B. 2 million.

    C. 5 million.

    D. 10 million.

2. The three most popular methods of birth control worldwide are

    A. the pill, IUDs, and rhythm method.

    B. the pill, sterilization, and rhythm method.

    C. sterilization, IUDs, and the pill.

    D. sterilization, the pill, and condoms.

3. The Chinese government estimates China's carrying capacity at about _____ people.

    A. 5 billion

    B. 2 billion

    C. 700 million

    D. 500 million

4. Excluding abstinence, the most effective method of preconception birth control is

    A. sterilization.

    B. condom with spermicide.

    C. diaphragm.

    D. IUD.

5. All of the following are postconception birth control methods except

   A. Depo-Provera.

   B. abortion.

   C. IUD.

   D. RU-486 pill.

Short Answer

1. What was the Mexico City policy?
2. What is family planning?
3. In the U.S., which two groups are least likely to use birth control?
4. What three factors contributed to lower population growth in the industrialized world around 1900?
5. What are the four stages of the demographic transition?

Thought Questions

Develop a complete answer for each of the following.

1. Discuss the growth of human population throughout history, beginning with early hunter-gatherers and moving through the onset of agriculture, the rise of cities, and the Industrial Revolution.
2. Explain the demographic trap being experienced by some developing countries. What measures would you propose to help nations escape from this trap?
3. Define population policy. What are indications that the U.S. has an unofficial pronatalist policy? Give examples of pronatalist and antinatalist policies from other countries.
4. Outline the arguments in favor of controlling or limiting population growth.
5. Explain how contraceptive use differs between LDCs and MDCs. Which are most effective in LDCs? In MDCs? Discuss the reasons for these differences.

> **Did You Know...?**
>
> In the last 100 years, except for a few sparsely populated oil-rich states, no poor country with high population growth has developed economically without first stabilizing its birth rate.

Suggested Activities

1. Think about your own views and those of your family or community: Do you believe population growth should be controlled? If so, what means should be used or implemented in order to control it? What values do you have that determine your answer to this question?

2. Choose a country that is foreign to you and about which you know very little. (If you are of European descent, choose a country in Asia, Africa, or South America; similarly, if you are of Asian, African or South American descent, choose a country located in a very different part of the world.) Discover what demographic conditions prevail in that country: Is it experiencing fast, slow, or no growth? What does its population profile look like? Is it economically developed or is it trying to develop? What are its major industries or markets? Imagine that you have been asked to help craft a population policy for this country. What would you recommend? If you suggest limiting growth, what means would you employ to achieve this goal? Keep in mind the nation's history, culture, and customs so that you can formulate an effective plan.

Chapter 10
Food Resources, Hunger, and Poverty

 **Chapter Outline**

I. **Describing Food Resources: Biological Characteristics**

   A. What Are the Components of a Healthy Diet?

   1. The Body Needs Food for Three Main Reasons

      a. Supply the energy needed to do work

      b. Grow and develop

      c. Regulate life processes

   2. Essential Elements for Life

      a. Macronutrients — required by the body in large amounts

         i. Carbohydrates and Fats (primary energy source to do work)

         ii. Proteins (the body's "building blocks")

      b. Micronutrients — required by the body in trace amounts

         i. Iron

         ii. Vitamins

   B. What Foods Do Humans Rely on?

   1. Staples — Important Foods in Many People's Lives

   2. Four Especially Important Food Crops: Wheat, Rice, Maize, Potato

   3. Nine Especially Important Animals: Cattle, Pigs, Sheep, Horses, Poultry, Mules/Asses, Goats, Camels, Buffalo

II. **Describing Food Resources: Physical Characteristics**

   A. Where Does Our Food Come From?

   1. Available Fertile Land

   2. Aquatic Harvest

   B. What is the Current Status of Food Production?

   1. Grain

      a. Consumed directly, provides half of human food energy intake

      b. Consumed indirectly (eggs, milk, meat), provides part of remainder

c. Good indicator of food security — ability of a nation to feed itself on an ongoing basis

d. Total production grew 4.6 percent per year between 1950 and 1990.

e. Since 1990, grain production has increased at an annual rate of 0.5 percent.

2. Fish

a. Leading source of animal protein worldwide

i. Primary source of protein for one billion people

ii. About one-third of catch is not used for human consumption

b. Since 1988, the marine catch per person has declined nine percent.

c. Eleven of the world's 15 most important ocean fisheries are harvested at or beyond maximum sustainable levels.

d. Decline in ocean fisheries linked to several causes

i. Overharvesting

ii. Habitat destruction

iii. Pollution

3. Meat

a. Total production rose to 232 million tons in 1997.

b. Per capita production has doubled since mid-century.

c. Pork accounts for largest share of world meat production, followed by beef, poultry, and mutton.

d. Poultry production rising fastest of all meats, largely because its conversion efficiency is much higher.

III. Describing Food Resources: Social Characteristics

A. What Is Hunger and How Does It Affect Human Health?

1. Starvation and Famine

a. Starvation is defined as suffering/death from deprivation of nourishment.

b. Famine is widespread starvation as a result of many factors.

i. Catastrophic natural events

ii. Human activity such as war or inequitable land distribution

2. Undernutrition and Malnutrition

a. Chronic undernutrition

- i. The consumption of too few calories/protein over an extended period of time
- ii. Occurs when a person regularly consumes fewer than 2,000 calories
    b. Malnutrition
        i. Generally defined as the consumption of too little of the specific nutrients essential for good health
        ii. Can maim and deform
    c. Malabsorptive hunger
        i. Often accompanies undernutrition and malnourishment
        ii. Body loses ability to absorb nutrients from food consumed
        iii. Caused by parasites in digestive tract or severe protein deficiency
  3. Nutritional Diseases
     a. Kwashiorkor
        i. Correct calorie count
        ii. Insufficient protein intake
     b. Marasmus
        i. Deficient calorie count
        ii. Deficient protein intake
  4. Seasonal Hunger
     a. Way of life — part of annual cycle for many
     b. Occurs at a time before the new harvest when old harvest runs out
     c. United States — end of month when food stamps run out
B. Who Are the Hungry and Where Do They Live?
C. What Is the Relationship between Hunger and Poverty?
   a. Hungry all share common characteristic of being poor.
   b. Everyday reality (implications) of poverty:
      i. Compare income with the price of buying food items
      ii. International Human Suffering Index
D. What Is the Relationship between Poverty and the Environment?
   a. Hungry forced to degrade environment to support themselves.
   b. Degraded environment offers little to support people, resulting in increased poverty.

E. Why Does Hunger Exist in a World of Plenty?

    a. Natural environmental conditions

    b. Lack of political will

    c. Various economic factors

        i. Best lands controlled by small numbers of wealthy landowners

        ii. Emphasis on cash crops for export

IV. Managing Food Resources

  A. How Did Humans Manipulate Food Resources in the Past?

    1. Domestication

      a. Result of selection of traits useful for survival in captivity

      b. Led to rise of agriculture

    2. Agriculture

      a. Purposeful tending of a particular plant/animal species for human use

      b. Land races — varieties of plants adapted for local conditions

  B. How Did Agriculture Change in the Twentieth Century?

    1. Developed World: Trend Toward Intensive Farming and High Inputs of Fertilizers, Herbicides, and Pesticides

      a. Advantages: greatly increased yields

      b. Disadvantages: expensive, fossil-fuel intensive, encouraged large monocultures

    2. Developing World: Green Revolution

      a. Advantages: High-yield varieties helped to double crop production.

      b. Disadvantages: High-yield varieties require significant inputs of fertilizers, irrigation water, and pesticides and so are expensive and may degrade the environment.

  C. How Can We Improve the Global Food Supply?

    1. Modify the Diet of People in the Developed World

    2. Improve Management of Ocean Fisheries

    3. Add Plants to Human Diet

      a. Research and discover "new" (little-used) plants that could supplement human diet

b. Gene bank — place where germ plasm of plant species is preserved for future use

4. Increase Yields Through Aquaculture — Production of Aquatic Plants/Animals in Controlled Environment

5. Increase Yields Through Biotechnology — Organisms or Their Parts Used to Manufacture Useful or Commercial Products

Learning Objectives

After learning the material in Chapter 10, you should be able to:

1. Identify the critical components of a healthy diet.
2. Identify the major foods relied on by the global human population and summarize the current status of food production.
3. Describe the various manifestations of hunger and explain how hunger affects human health.
4. Discuss how food consumption patterns vary worldwide.
5. Describe the relationship among hunger, poverty, and environmental degradation.
6. List various reasons why hunger continues to be a problem.
7. Briefly describe how agriculture has changed in the past fifty years in both the developed and the developing worlds.
8. Explain the strategies that could be used to increase the global food supply.

Did You Know...?

You can produce over 1,000 cherry tomatoes in four square feet of land!

Key Concepts

Read this summary of Chapter 10 and identify the important concepts discussed in the chapter.

Carbohydrates and fats are the major sources of the energy required to maintain the body and perform work. Proteins are the "building blocks" of the body, forming the substance of muscles, organs, antibodies, and enzymes.

Of the 80,000 potentially edible crops on Earth, global agriculture is dependent on only a few plant species. Just eight crops supply 75 percent of the human diet, with the top four being wheat, rice, corn (maize), and potatoes. Agriculture also relies on just a small number of animals. Meat production can cause problems when it is practiced to the extent and in the manner common in the developed world. When livestock are fed grains, the efficiency of food production is greatly reduced.

Three important food production sectors are grain, fish, and meat. Grain directly supplies about half of human food energy intake and indirectly supplies part of the remainder. It is a good indicator of food security, the ability of a nation to feed itself on an ongoing basis. Building and maintaining carryover stocks, the total amount of grain in storage when the new harvest begins, is becoming increasingly more difficult. Per capita production fell seven percent between 1984 and 1998, and in 1996, carryover stocks equaled just 50 days of consumption — the lowest level ever record.

Fish is the leading source of animal protein worldwide and is especially important in Asia, the coastal nations of Africa, and island nations. As the harvesting and export of fish rises, and fish consumption in wealthy nations increases, the poor find it more difficult to catch or purchase the fish they depend upon. Pollution and habitat destruction have degraded or destroyed some fisheries, but the chief culprit in declining stocks is overharvesting. The per capita marine catch (which accounts for the lion's share of the total world fish catch) has fallen nine percent since 1988. Eleven of the world's 15 most important ocean fisheries are currently harvested at or beyond maximum sustainable levels.

In terms of share of total world meat production, the four most important meats are pork, poultry, beef, and mutton. Total world meat production continues to rise slowly and has managed to keep pace with population growth, but trends vary according to the type of meat. The rate of increase in beef production has slowed in recent decades; total production of mutton has risen only slightly since mid-century. In contrast, pork and poultry production have both increased rapidly, and these meats are the two most important contributors to the human diet.

Most of the world's hungry live in LDCs. The most extreme and dramatic effect of hunger is starvation, suffering or death from the deprivation of nourishment. Widespread starvation, or famine, such as that which occurred in Somalia in 1992, is the result of many interrelated factors, including war, prolonged drought, production of cash crops, unequal distribution systems, and unequal distribution of lands. Each year, however, of the deaths attributed to hunger, most come not from starvation but from undernutrition, the consumption of too few calories and proteins over an extended period of time; malnutrition, consumption of too little (or infrequently, too much) of specific nutrients essential for good health; malabsorptive hunger, the body's loss of the ability to absorb nutrients from the food consumed; and hunger-related diseases, such as kwashiorkor and marasmus.

Worldwide, chronic hunger affects about one billion people. About 40 percent of the world's hungry are children; most of the rest are women. Most of the world's hungry live in the developing world within the "great hunger belt," a region encompassing various nations in Southeast Asia, the Indian subcontinent, the Middle East, Africa, and the equatorial region of Latin America. Five nations — India, Bangladesh, Nigeria, Pakistan, and Indonesia — together house one-half of the world's hungry. It is important to remember, however, that even within the wealthiest industrial nations, some people go hungry. No matter where they live, the hungry share a common characteristic: poverty.

There is a very real connection between hunger, poverty, and environmental degradation. Although much of the world's environmental degradation is due to affluence in MDCs, poverty, especially in LDCs, also causes environmental degradation. The poor are forced to cultivate marginal lands, overgraze grasslands and pastures, or clear forests. Hunger and poverty trap the poor in a vicious cycle. Given less food and nutrients, the poor are less able to work and learn and more prone to disease; lack of education or poor work performance prolongs or exacerbates poverty, as does chronic illness. For many, death is the only way out of the cycle of hunger and poverty.

Food production, especially of grains, is sufficient to feed the world's people, yet hunger persists in our world. Its causes are many: natural environmental conditions which account for disparities in food production; economic and political institutions which concentrate power so that some people are left with none; inequalities in land ownership; and cash crop production.

In the past fifty years, food production in the United States has come to rely heavily on inputs of energy, water, synthetic fertilizers, herbicides, and pesticides. In the developing world, the green revolution, sparked by the use of high-yield varieties of crops, promised to make nations self-sufficient in food production. But despite its successes worldwide, the green revolution has not ended hunger in the developing world. In recent years, the international community has begun several efforts to protect genetic diversity (through gene banking) and increase world food production by modifying the diet of people in the developed world, improving the management of ocean fisheries, adding new plants to the human diet, and increasing our use of aquaculture and biotechnology).

Key Terms

aquaculture	agriculture
amino acids	biotechnology
carryover stocks	cash crops
centers of diversity	chronic undernutrition
cryopreservation	domestication
estuary	famine
food security	gene bank
genetic engineering	green revolution
hunger	kwashiorkor
land races	malabsorptive hunger

malnutrition

monoculture

seasonal hunger

starvation

marasmus

polyculture

staples

tissue culture

Environmental Success Story

In Austin, Texas, elementary students are being served by the Green Classroom program. Since 1989, the Green Classroom has educated low-income students in inner-city elementary schools about environmental issues. Each week, students teach mini-courses in composting, organic gardening, water quality, conservation studies, and energy conservation to other students in the district. They also grow and market organic produce, work with senior citizens, and participate in cooking and nutrition classes.

Source: The Green Classroom. "Seventh Annual National Awards for Environmental Sustainability Winners: Food and Agriculture." *Renew America* (22 January 1997). Available: http://solstice.crest.org/environment/renew_america/winners.html. 18 February 1999.

True/False

1. The nine essential amino acids are those that are synthesized by the body.   T   F

2. More land is used for grazing livestock than for growing crops.   T   F

3. On average, people in MDCs consume about 10 percent more calories than necessary.   T   F

4. Chronic hunger affects approximately one billion people worldwide.   T   F

5. Aquaculture is the fastest growing sector of agriculture in the U.S.   T   F

Fill in the Blank

1. _____ and _____ are the major sources of energy for the body.

2. Over one billion people in Asia, the coastal nations of Africa, and island nations depend upon _____ as their primary source of animal protein.

3. _____ is a type of hunger that can be caused by parasites in the intestinal tract.

4. Varieties of plants that are adapted to local conditions are called _____.

5. The process of storing plant or animal tissue in liquid nitrogen is called _____.

Multiple Choice

Choose the best answer.

1. Complex carbohydrates, protein, and fat are examples of

    A. micronutrients.

    B. macronutrients.

    C. staples.

    D. None of the above is true.

2. At most, the portion of ice-free land that could be used to grow crops is about

    A. 11 percent.

    B. 22 percent.

    C. 33 percent.

    D. None of the above is true.

3. Worldwide, the number of people who die each year from hunger and hunger-related diseases is estimated to be around

    A. 5 million.

    B. 15 million.

    C. 20 million.

    D. 40 million.

4. A diseased caused by a lack of protein and too few calories is

    A. marasmus.

    B. kwashiorkor.

    C. malabsorptive hunger.

    D. amaranth.

5. The number of potentially edible crops on Earth is estimated to be

    A. 25,000.

    B. 50,000.

    C. 80,000.

    D. None of the above is true.

Short Answer

1. Why are proteins important for a healthy diet?

2. What are the top four food crops in the world?

3. What is famine?

4. What is the "great hunger belt"?

5. What is the International Human Suffering Index and what does it reveal about a country? What are demographic conditions like in countries that have the greatest human suffering?

Thought Questions

Develop a complete answer for each of the following.

1. Describe a healthy diet.

2. What kinds of hunger are there? Explain the situations that lead to each.

3. Explain the energy investment required to support a diet with a high meat content. Discuss the fact that people in MDCs consume twice as much meat as people in LDCs.

4. Where does hunger occur? Compare the diets of people in MDCs and LDCs and of wealthy people and poor people.

5. Why does hunger exist?

6. Compare pre-1950 agriculture in the developed world with agriculture today. Explain why, even though grain production per hectare has increased, there are more hungry people than there were ten years ago. How have high yield varieties and cash crops contributed to the situation?

7. Describe the various techniques being used to preserve genetic diversity of crop plants. Why is this work important?

▪ Related Concepts

Describe the relationship. (There may be more than one.)

BETWEEN...	AND...
malnutrition	undernutrition
hunger in LDCs	hunger in MDCs
Human Suffering Index	poverty
poverty	environmental degradation
democracy	hunger

Did You Know...?

One-quarter of the grapes eaten in the U.S. are grown 7,000 miles away, in Chile.

 ## Suggested Activities

1. Keep a "food diary" of everything you eat each day for one week. How many calories do you consume? What percentage of calories come from fat, protein, and carbohydrates?

2. Find out more about hunger in your area. Who are the hungry? Why do they remain hungry in a world of plenty?

3. Try an experiment: for two weeks, modify your diet to include more foods that are produced locally.

Chapter 11
Energy Issues

 **Chapter Outline**

I. **Describing Energy Resources: Physical Characteristics**

   A. What Is Energy and How Is It Measured?

   1. Ability To Do Work

      a. Kinetic

      b. Potential

   2. Measurements

      a. Quantity of different forms of energy

      b. Amount of specific types of energy resources

   B. How Are Energy Resources Classified?

   1. Nonrenewable

      a. Exist in a finite supply or are renewable at a rate slower than the rate of consumption

   2. Renewable

      a. Resupplied at rates faster than or consistent with use

   3. Perpetual

      a. Originate from a source that is virtually inexhaustible

   4. Alternative

      a. Perpetual and renewable resources that offer a choice beyond fossil fuels

   C. What Is Energy Efficiency?

   1. Measure of the percentage of the total energy input that does useful work and that is not converted into diffuse, excess heat

   2. Net Efficiency

   3. Life-Cycle Cost

II. **Describing Energy Resources: Biological Characteristics**

   A. Many energy resources, including all fossil fuels, are derived from organic matter.

   B. What Is the Biological Significance of Energy Resources?

1. Every species on the planet requires energy for growth, maintenance, and reproduction.

III. Describing Energy Resources: Social Characteristics

A. Most human societies are possible only by the capture, conversion, and use of enormous amounts of energy.

B. How Do Present Energy Consumption Patterns Vary Worldwide?

C. What Important Environmental Issues Are Related to the Use Of Energy Resources?

1. Social Changes
2. Environmental And Health Effects
3. Dependence On Fossil Fuels
4. Nuclear Power
5. Energy Policy

Learning Objectives

After learning the material in Chapter 11, you should be able to:

1. Define energy and explain how it is measured.
2. Explain how energy resources are classified.
3. Define energy efficiency and summarize its environmental and economic benefits.
4. Describe how energy consumption patterns vary worldwide.
5. Identify five broad issues related to energy consumption.

Did You Know...?

Making a standard disposable battery uses 50 times the power it generates.

Key Concepts

Read this summary of Chapter 11 and identify the important concepts discussed in the chapter.

Energy, the ability to do work, is part of the abiota, the nonliving or physical component of the biosphere. All energy is either kinetic — energy due to motion or movement — or potential — energy in storage, that is, energy which can be converted to another form. Energy can take different forms, such as heat, light, chemical, or electrical energy. Various units are used to measure the *quantity* of different *forms* of energy. For example, heat energy can be measured in calories, British thermal units (btu), or therms; electrical energy is measured in kilowatts. In addition to measuring the quantity of a specific form of energy, such as heat, we also measure the *amount* of a specific type of energy *resource*. Petroleum, for example, is measured in barrels, while natural gas is measured in cubic feet and coal is measured in short tons.

Energy resources are generally classified as nonrenewable, renewable, or perpetual. Nonrenewable resources exist in finite supply or are renewed at rates far slower than the rate of consumption. Renewable resources are resupplied at rates faster than is consistent with use. Perpetual resources are virtually inexhaustible, at least in time as measured by humans. Renewable and perpetual resources are called alternative resources because they offer us an alternative to fossil fuels.

In any energy conversion, although the same amount of energy exists before and after the conversion, not all of the energy remains useful. The conversion of energy from one form to another form always involves a change or degradation from a higher quality form to a lower quality form, and all forms of energy are ultimately degraded to heat energy. Energy efficiency is a measure of the percentage of the total energy input that does useful work and is not converted to low temperature, low-quality heat; it can determine the efficiency of any system, device, or process that converts energy. The net efficiency of a process or system that includes two or more energy conversions is found by determining the efficiency of each conversion. Efficiency is an important factor to bear in mind when purchasing an electrical appliance or gadget. The life-cycle cost of an item includes both the initial cost plus the operating costs, which will be lower for more efficient items.

Most of the world's energy is used by the more-developed countries (MDCs); just 20 percent of the global population uses 70 percent of the energy. Energy issues are complex and involve all aspects of culture.

Energy issues can be grouped into five broad categories: social changes, environmental and health effects, dependence on fossil fuels, nuclear power, and energy policy. With respect to social changes, per capita consumption of energy in MDCs is currently six times higher than per capita consumption in less-developed countries (LDCs), but LDCs are outpacing MDCs in terms of growth of energy consumption. As LDCs continue to attempt economic development, energy supplies will become more precious and more threatened.

All energy consumption affects the environment and human health. The type and severity of these effects differ significantly according to the resource used. In general, renewable and perpetual resources are far more benign than fossil fuels.

Despite the environmental problems they cause and their low conversion efficiency, fossil fuels are the lifeblood of industrial societies. However, our decades-long reliance on fossil fuels has greatly diminished supplies.

The debate over nuclear energy pits its proponents — who argue that nuclear energy is a clean, safe resource which can help reduce emissions of greenhouse gases — against those who argue that it is inherently unsafe and that its drawbacks (accidents, dismantling of old plants, and disposal of nuclear wastes) are too severe to continue or increase our reliance on this form of energy.

In the past two decades, and especially in the 1990s, Americans have enjoyed a steady and ample supply of inexpensive energy that sustains their highly consumptive lifestyles. Though the nation could use this time of plenty to develop long-term, cleaner energy resources, the U.S. National Energy Policy Plan encourages continued reliance on fossil fuels and increased production of domestic oil reserves. Supplies of petroleum and natural gas are likely to become virtually depleted by the middle of the twenty-first century, but through early planning the United States can develop and implement a national energy policy that is both effective and farsighted.

 Key Terms

alternative resource	energy
energy efficiency	kinetic energy
life-cycle cost	net efficiency
nonrenewable resource	perpetual resource
potential energy	renewable resource

Environmental Success Story

Who says environmentalism is hard to stomach? Not the creators of the Rainforest Cafe, an environmentally sound restaurant that opened in the fall of 1994 in Minnesota's Mall of America and now has expanded to 22 domestic locations and eight international restaurants. This unique eatery offers 130 types of fish, along with various other reptiles and unusual critters. Food is chosen carefully to avoid certain practices or businesses which harm the environment. Everything about this cafe chain is environmentally sound — right down to the table tops, which are compiled of recycled newspapers and soybean-based resin.

Source: "Rainforest Cafe Wins Top Honors From Industry Leaders." *Rainforest Cafe Website.* (3 March 1999). Available: http://www.rainforestcafe.com/main.html/. 17 March 1999.

True/False

1. The sun is an example of a renewable resource.   T   F

2. Renewable and perpetual resources are called alternative resources.   T   F

3. Per capita energy consumption in MDCs is six times greater than that in LDCs. T F

4. The construction of nuclear power plants has steadily increased in recent years. T F

5. The 1998 National Energy Policy Plan primarily seeks to fund research for renewable, energy-efficient technologies. T F

Fill in the Blank

1. Fossil fuels are also called _____ because they exist in a finite supply.

2. _____ resources are supplied at rates faster than or consistent with use.

3. _____ is a measure of the percentage of the total energy input that does useful work and is not converted into low temperature, low-quality heat.

4. The initial cost of an energy-efficient device plus the lifetime operating cost is called the _____.

5. The unofficial energy policy of the U.S. government has been to _____.

Multiple Choice

Choose the best answer.

1. Solar energy is an example of a

    A. nonrenewable resource.

    B. conventional resource.

    C. renewable resource.

    D. perpetual resource.

2. The U.S., with roughly five percent of the world's people uses _____ of the world's fossil fuels.

    A. 5 percent

    B. 15 percent

    C. 25 percent

    D 50 percent

3. Alternative energy sources provide about _____ of the world's energy.

    A. 5 percent

    B. 10 percent

    C. 20 percent

    D. 30 percent

4. Which of the following is true?

    A. Per capita energy consumption in MDCs is greater than that in LDCs.

    B. Energy consumption is growing faster in LDCs than in MDCs.

    C. Energy consumption is increasing faster than the population in both MDCs and LDCs.

    D. All of the above are true.

5. At the current rate of U.S. energy consumption, by the year 2010, our dependence on foreign oil will be _____ percent.

    A. 40 percent

    B. 60 percent

    C. 70 percent

    D. 90 percent

Short Answer

1. Why do lightbulbs get hot?

2. What is life-cycle cost?

3. What is energy efficiency?

4. What is an energy policy?

5. What is the first law of thermodynamics?

Thought Questions

Develop a complete answer for each of the following.

1. Explain the concept of net efficiency and discuss how it affects the economic and environmental costs of producing electricity.

2. Compare energy use in MDCs with energy use in LDCs.

3. Discuss the social changes that will be necessary if we are to conserve nonrenewable resources for uses other than supplying energy.

4. What are the implications of MDCs' dependence on fossil fuels?

5. Present the arguments for and against nuclear power.

6. What are the consequences of the United States's current National Energy Policy Plan?

7. Discuss how the areas of major energy issues presented in this chapter — social changes, environmental effects, dependence on fossil fuels, nuclear power, and energy policy — are interconnected.

▢▨▪ Related Concepts

Describe the relationship. (There may be more than one.)

BETWEEN...	AND...
perpetual resource	renewable resource
nonrenewable resource	conventional resource
energy conservation	economic development
energy policy	"drain America first"

Did You Know....?

One barrel of crude oil provides 2.5 quarts of virgin motor oil, while one gallon of recycled motor oil yields the same number of quarts of high quality motor oil.

 ## Suggested Activities

1. Find out where the energy you use in your home comes from.

2. Organize a classroom debate on one or more of the energy issues presented in this chapter.

3. List the appliances in your home. If your energy supply were suddenly stopped, how would you manage without each appliance?

4. Try an experiment. Spend an entire day without using electricity or other types of energy. Share your experience with your classmates.

Chapter 12
Energy: Fossil Fuels

 Chapter Outline

I. **Describing Fossil Fuel Resources: Biological Characteristics**

   A. What Are Fossil Fuels?

      1. Coal, Petroleum, Natural Gas, Oil Shales, and Tar Sands

      2. Composed Chiefly of Carbon and Hydrogen

   B. How Were Fossil Fuels Formed?

II. **Describing Fossil Fuel Resources: Physical Characteristics**

   A. Where Are Fossil Fuel Deposits Located?

      1. Proven or Economic Reserves

      2. Subeconomic Reserves

      3. Indicated or Inferred Reserves

III. **Describing Fossil Fuel Resources: Social Characteristics**

   A. What Is the Current Status of the World's Major Fossil Fuels?

      1. Coal

         a. Anthracite

         b. Bituminous

         c. Subbituminous and Lignite

      2. Petroleum

      3. Natural Gas

         a. Associated Gas

         b. Nonassociated Gas

      4. Minor Fossil Fuels: Oil Shales and Tar Sands

IV. **Managing Fossil Fuel Resources**

   A. How Have Fossil Fuels Been Used Historically?

   B. How Has Energy Consumption Changed in the United States?

      1. Wood was the primary fuel for the early pioneers of the United States, but gradually coal, followed by gasoline, became the primary source of energy.

2. Since 1960, oil, coal, and natural gas have played dominant roles in U.S. energy consumption.

C. What Was the 1973 Oil Embargo?

1. The oil embargo caused an "energy crisis" in the U.S. and forced Americans to think realistically about their energy consumption — a conservation revolution.

D. How Can We Increase Energy Efficiency?

1. Industries
2. Communities
3. Individuals

Learning Objectives

After learning the material in Chapter 12, you should be able to:

1. Identify the three major types of fossil fuels and explain briefly how fossil fuels were formed.

2. Describe the current status (uses, availability, and environmental effects) of coal, petroleum, natural gas, oil shales, and tar sands.

3. Summarize how energy consumption has changed in the U.S. in the past 200 years.

4. Give examples to illustrate how industries, communities, and individuals can increase energy efficiency and conserve fossil fuels.

Did You Know...?

The average fuel economy of cars has dropped each year since 1988, and is now at 27.5 miles per gallon.

 ## Key Concepts

Read this summary of Chapter 12 and identify the important concepts discussed in the chapter.

Coal, petroleum, and natural gas, the world's conventional fuels, are the fossilized remains of organic matter. Coal originated from plant matter, petroleum from aquatic organisms (algae and plankton), and natural gas from all types of organic matter, even cellulose. The processes that formed fossil fuels continue today, but because those processes are extremely slow, fossil fuels are classified as nonrenewable and finite. Fossil fuels are not distributed evenly beneath the Earth's surface because the conditions that gave rise to the preservation and fossilization of organic matter did not occur everywhere.

The major fossil fuels are coal, petroleum, and natural gas. Minor fossil fuels are oil shales and tar sands. Coal, a solid composed primarily of carbon with small amounts of hydrogen, nitrogen, and sulfur, is the most abundant fossil fuel on Earth. There are four types of coal. Anthracite coal has the highest carbon content and the lowest sulfur content; it is the most efficient and cleanest burning type of coal and the most preferred for heating homes and commercial buildings, but it is also the least abundant. Bituminous coal, the most common type, accounts for over half of U.S. reserves and has a heating value slightly lower than that of anthracite. It is preferred for electric power generation and the production of coke, which is used to make steel. Subbituminous and lignite coals, which together account for a little less than half of U.S. reserves, both have low heating values and must be burned in large amounts in order to heat effectively. However, these types contain very little sulfur, a desirable trait since sulfur emissions contribute to the formation of acid precipitation. Coal consumption declined over the past century but is expected to increase as reserves of petroleum dwindle and as reliance on nuclear power continues to slow. The majority of the world's coal reserves are located in the United States, the former Soviet Union, and China. U.S. reserves lie in three broad regions: Appalachian, Interior, and western or Rocky Mountain. Wyoming, West Virginia, and Kentucky are the three highest coal-producing states. Coal has the most serious environmental and health effects of all energy resources in use today. Problems include acid precipitation, acid mine drainage, air and water pollution, and carbon dioxide emissions, which exacerbate global climate change.

Petroleum, or crude oil, is a liquid composed primarily of hydrocarbon compounds. It is perhaps the most versatile of all fossil fuels, used to make products such as propane, gasoline, jet fuel, road tar, motor oil, and thousands of non-fuel substances and chemicals. The United States accounts for about 25 percent of the annual global consumption of petroleum (1,050 gallons per person, per year). Gasoline is the most consumed petroleum product in the United States; the transportation sector accounts for about 60 percent of the nation's annual oil consumption. The greatest reserves of petroleum lie in the Middle East; other areas with sizable deposits include Latin America, the former Soviet Union, Africa, and North America. Like coal, petroleum has serious environmental consequences, including air pollution (its combustion releases carbon monoxide, sulfur oxides, nitrogen oxides, and hydrocarbons). Oil spills pose another serious threat to the environment.

Several different types of gas comprise natural gas; the most abundant is methane. Natural gas usually occurs along with petroleum and is known as associated gas. Increased attention has focused on natural gas because of our growing awareness of finite petroleum supplies and because gas is a much cleaner burning fuel than oil or coal. Compared to coal or petroleum, the combustion of natural gas releases far less carbon dioxide, nitrogen oxides, and sulfur oxides. Natural gas systems are also desirable because of their high conversion efficiency. Natural gas is a fairly abundant fuel; the world's largest deposits are in the former Soviet Union and the Middle East. In the United States, gas production is centered around the Gulf Coast, the mid-continent, and west Texas-east New Mexico.

Oil shales, fine-grained, compacted sedimentary rocks that contain varying amounts of a waxy, combustible organic matter called kerogen, and tar sands, sandstones that contain bitumen, a thick, high-sulfur, tarlike liquid, are minor fossil fuels. Extracting and refining these fuels are expensive and energy-intensive processes. Extraction also disturbs large areas of land, requires significant amounts of water, and produces a good deal of waste rock.

Humans have used fossil fuels for various purposes for thousands of years. Use as a fuel is relatively recent, at least for petroleum and natural gas. In the United States, wood was the leading fuel source for some two hundred years but was displaced by coal by 1885. In the early 1900s, coal met over 90 percent of the nation's energy needs. Coal's dominance ended with the widespread use of petroleum. By 1946, natural gas and petroleum together displaced coal as the chief energy source; by 1950, petroleum consumption alone outpaced coal consumption. In 1947, the United States became a net petroleum importer, a condition that persists today. The 1973 embargo by the Organization of Petroleum Exporting Countries curtailed supplies, and the price of oil skyrocketed. An increased emphasis on conservation and energy efficiency resulted, but these were short-lived. Prices have since fallen, despite the occasional fear of shortages (such as that due to the Persian Gulf War), and conservation efforts have waned. But the world's supplies of fossil fuels are finite, and as supplies become increasingly scarce, their price will rise significantly. Increasingly, our reliance on fossil fuels will become economically and environmentally costly.

 ## Key Terms

acid drainage	anthracite
associated gas	bitumen
bituminous	coal
conservation revolution	demand side management
economic reserves	fossil fuels
indicated reserves	inferred reserves
land subsidence	lignite
natural gas	nonassociated gas
oil shale	overburden
petroleum	proven reserve
subbituminous	subeconomic reserves
tar sands	

Environmental Success Story

In Thousand Palms, California, the SunLine Transit Agency had to replace its entire fleet of buses in the late 1990s. The replacement was revolutionary as SunLine became the first transit authority to completely switch to vehicles that run on compressed natural gas (CNG). This gas is better than gasoline because it burns cleaner, costs less, and produces fewer greenhouse gases. The move also seems to have increased their ridership by 11 percent and has inspired the local post office to switch to CNG for their 35 vehicles. SunLine has put more than 100 CNG-powered vehicles on the road in its area.

Source: SunLine Transit Agency. "Eighth Annual National Awards for Environmental Sustainability Winners: SunLine CNG." *Renew America* (22 April 1998). Available: http://solstice.crest.org/sustainable/renew_america/winner98.html. 24 March 1999.

True/False

1. Coal and petroleum originated largely from plant material.   T   F
2. Natural gas is the most abundant fossil fuel on Earth.   T   F
3. Petroleum is primarily composed of a mixture of oxygen, sulfur, and nitrogen compounds.   T   F
4. Annual global usage of petroleum equals about 27 billion barrels.   T   F
5. Natural gas is the cleanest of all fossil fuels.   T   F

Fill in the Blank

1. The era in which most of the world's coal was formed was about _____ million years ago in the _____ period.
2. The energy contained in fossil fuels came originally from _____, captured by plants through _____.
3. The petroleum deposits that have been located, measured, and inventoried are called _____.
4. The tone of coal that burns most efficiently is _____.
5. _____ is a new approach being used by some utilities that encourages energy efficiency and conservation by electricity consumers, both residential and industrial.

Multiple Choice

Choose the best answer.

1. Of what two elements are fossil fuels primarily made?

    A. carbon and oxygen

    B. hydrogen and oxygen

    C. carbon and hydrogen

    D. carbon and nitrogen

2. The most common type of coal in the U.S. is

    A. anthracite.

    B. bituminous.

    C. lignite.

    D. subbituminous.

3. U.S. coal reserves represent about _____ of the United States's total energy resources.

    A. 50 percent

    B. 60 percent

    C. 80 percent

    D. 90 percent

4. The most productive coal-producing state is

    A. Montana.

    B. Kentucky.

    C. West Virginia.

    D. Wyoming.

5. Natural gas

    A. is primarily made up of methane, ethane, and propane.

    B. has about a 50 percent energy efficiency.

    C. is mostly used in the residential sector of the U.S.

    D. All of the above are true.

Short Answer

1. How is coal formed?
2. How is petroleum formed?
3. How is natural gas formed?
4. How many products are created from petroleum? Which is the most heavily consumed in the U.S.?
5. What is coke?

Thought Questions

Develop a complete answer for each of the following.

1. Explain why predictions and estimates of remaining fossil fuels are not completely reliable.
2. Discuss the environmental impacts of the production and use of coal.
3. Discuss the environmental impacts of the production and use of petroleum.
4. Compare the environmental impacts of natural gas with those of other fossil fuels, including tar sands and oil shales.
5. What are the major uses of petroleum? If the U.S. suddenly had its supply of oil cut in half, what uses would you eliminate (if any), and why?
6. Why have oil shales and tar sands played only a minor role as energy sources? Might this change in the future? Why or why not?
7. How have humans used fossil fuels throughout history?

▢▨■ Related Concepts

Describe the relationship. (There may be more than one.)

BETWEEN...	AND...
petroleum	oil shale
surface mining	acid drainage
petroleum	natural gas
OPEC	oil shortages of the 1970s
energy efficiency	CAFE standards

Did You Know...?

In 1997, the amount of carbon dioxide in the atmosphere reached its highest point in 160,000 years.

Suggested Activities

1. Read the *Wall Street Journal* for several weeks to follow gas prices and events in the world. What can you observe about how one influences the other?

2. Develop an energy conservation program for your community or campus. Include incentives to encourage people to participate.

3. Interview friends and family members about the 1973 OPEC oil embargo. What habits did they change? How did they feel about the situation?

4. Participate in an existing recycling program or start your own. Find out how plastics (made from petroleum) are recycled. Recycle used motor oil.

5. If possible, begin walking, riding a bike, or taking public transportation to school or work instead of driving a car.

Chapter 13
Energy: Alternative Sources

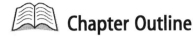 **Chapter Outline**

I. **Describing Alternative Energy Sources: Physical Characteristics**

   A. What Is Nuclear Energy?

      1. Energy contained within the nucleus of an atom, released by the fission of isotopes

      2. Present and Future Use

         a. Accounts for approximately 17 percent of the world's electricity and seven percent of total energy

         b. Accounts for about 20 percent of electric production in the U.S.

      3. Advantages and Disadvantages

         a. Advantages: does not emit carbon dioxide, sulfur oxides, or nitrous oxides

         b. Disadvantages: chance of accidents, difficulty of safe disposal of nuclear wastes, expensive

   B. What Is Solar Energy?

      1. Energy radiated from the sun to Earth

      2. Present and Future Use

         a. Passive solar systems

         b. Active solar systems

         c. Solar ponds

         d. Photovoltaic cells

      3. Advantages and Disadvantages

         a. Advantages: perpetual, non-polluting

         b. Disadvantages: not applicable in all areas, difficult to collect and store, high initial costs

   C. What Is Wind Power?

      1. Flowing rivers of air that provide a safe, clean source of perpetual energy

      2. Present and Future Use

         a. Second fastest growing source of electric power

3. Advantages and Disadvantages

   a. Advantages: perpetual, non-polluting

   b. Disadvantages: unreliable and intermittent, bird kills, aesthetic considerations

D. What Is Hydropower?

   1. Energy of falling water

   2. Present and Future Use

      a. Generates almost 23 percent of the world's electricity

   3. Advantages and Disadvantages

      a. Advantages: multipurpose, does not produce carbon dioxide

      b. Disadvantages: radically alters surrounding ecosystems, displaces human and nonhuman inhabitants

E. What Is Geothermal Power?

   1. Heat generated by natural processes occurring beneath the Earth's surface

   2. Present and Future Use

      a. Twenty-one countries produce 8,000 megawatts of electricity from geothermal resources — enough to heat over two million homes in cold climates.

   3. Advantages and Disadvantages

      a. Advantages: more environmentally sound than fossil fuels or nuclear energy

      b. Disadvantages: few deposits, primarily nonrenewable resource

F. What Is Ocean Power?

   1. Energy that can be derived from the seas

   2. Present and Future Use

      a. Tidal electric power plants

      b. Wave energy plants

      c. Ocean thermal energy conversion (OTEC)

   2. Advantages and Disadvantages

      a. Advantages: renewable, non-polluting

      b. Disadvantages: potential to reduce tidal range, reduce tidal current flow, alter sea levels, and cause death of migratory fish species

II. Describing Alternative Energy Sources: Biological Characteristics

A. What Is Biomass Energy?

1. Gaseous, liquid, or solid fuels derived directly or indirectly from plant photosynthesis

2. Present and Future Use

    a. Provides over 14 percent of all energy consumed worldwide

    b. Crop residues

    c. Alcohol fuels (ethanol)

    d. Gases released from decaying plant matter and animal waste (biogas)

3. Advantages and Disadvantages

    a. Advantages: accessible, inexpensive, clean

    b. Disadvantages: threat to land fertility

III. Describing Alternative Energy Sources: Social Characteristics

A. What Is Solid Waste?

1. Material that is rejected or discarded as being spent, worthless, useless, or in excess

2. Present and Future Use

    a. Shred and pelletize noncombustible materials to use as fuel

    b. Install biogas digesters in landfills

3. Advantages and Disadvantages

    a. Advantages: reduces amount of materials going to landfills

    b. Disadvantages: emissions of mercury and lead from incineration, hampers recycling efforts, questionable economic feasibility

IV. Managing Alternative Energy Sources

A. For centuries, people harnessed the power of wood, water, sun, and heat from the Earth.

B. The early twentieth century marked the beginning of the age of fossil fuels.

C. At the close of the twenieth century, our knowledge of diminishing fossil fuel supplies renews popular interest in perpetual and renewable resources.

Learning Objectives

After learning the material in Chapter 13, you should be able to:

1. Identify eight alternative energy resources, summarize their current use and future use, and explain the advantages and disadvantages of each.
2. Briefly summarize the role that alternative energy resources have played historically.
3. Describe how alternative energy resources might be developed to complement and eventually supplant fossil fuels.

Did You Know...?

Solar electric cells are used to power call boxes on the San Diego freeway.

Key Concepts

Read this summary of Chapter 13 and identify the important concepts discussed in the chapter.

Eight energy resources may prove to be especially important alternatives to fossil fuels: nuclear, solar, wind, hydropower, geothermal, ocean, biomass, and solid waste.

Nuclear energy, the energy contained within the nucleus of the atom, is released when an atom splits, or fissions, thus breaking into smaller atoms, releasing neutrons, and emitting heat energy. If the released neutrons bombard other fissionable atoms, a chain reaction results. Nuclear reactors are designed to sustain the fissioning process; they are powered by uranium-235 (U-235), a relatively rare fissionable uranium isotope (a form of an element that has the same number of protons but a different number of neutrons). A nuclear reactor core, housed within a containment vessel, heats water to produce steam that runs a turbine. While nuclear energy accounts for a modest proportion of the world's total energy supply (seven percent), its share of electric generation is significant — about 17 percent. France, Belgium, and several other countries rely on nuclear energy to produce as much as two-thirds of their electricity. The significant growth in the U.S. nuclear industry during the 1970s and early 1980s has tapered off in recent years, primarily due to safety concerns and less-than-anticipated demand for electricity.

The sun, source of the energy vital for life on Earth, can also provide energy for other purposes. Solar energy is used primarily for space and water heating. Passive solar systems incorporate design features of buildings and homes to capture the maximum amount of radiation from the sun in winter months and a minimum amount in the summer months. Only natural forces are used to distribute the heat. Active systems use fans or pumps driven by electricity to enhance the collection and distribution of the sun's heat. Solar energy may someday power electric generation on a significant basis. Photovoltaic cells generate electricity directly from sunlight. Atoms within these semiconductors absorb sunlight energy and liberate electrons, producing a direct electrical current. Solar power is advantageous because it exists everywhere (although intermittently in some areas), is non-polluting, conserves natural resources, and is technologically available for widespread use. However, it has several disadvantages: It is diffuse and so has to be collected over large areas to make it practical to use; it is also intermittent and therefore requires some means of storage. Moreover, the initial cost of solar power for a home or small business can be quite high (depending on the technology), although solar energy is less expensive than conventional fuels when life-cycle costs are determined.

The sun is also responsible for wind power. Because the sun warms different areas of the Earth and the atmosphere in unequal amounts, regional pressure differences result which produce winds. Historically, wind energy was a fairly significant source of power. It is enjoying renewed interest because it is a safe, clean source of perpetual energy. Wind turbines are often clustered in favorable geographic locations to make up wind farms. Disadvantages of wind power include the fact that it is variable and that it is site specific — it must be exploited in those areas where winds are reliable and strong enough to make the venture profitable (that is, areas where wind speed averages a minimum of 13 miles per hour).

Hydroelectric power results when water flows from the top of a dam to the bottom and strikes the blades of a turbine which drives an electric generator. Because most of the best sites for large-scale hydropower have already been developed, increased emphasis is being put on developing smaller sites that produce less electricity but are more environmentally sound. Hydropower generates almost one-fourth of the world's electricity, a contribution greater than that of nuclear power. Worldwide, it is responsible for the greatest proportion of electrical generating capacity of all renewable sources of energy. One advantage of hydropower is that it is non-polluting; in addition, dams are multipurpose, also providing for recreational facilities, municipal water supplies, irrigation, and flood control. On the negative side, dams destroy free-flowing rivers, flood lands, displace local residents, ruin local streams and waterfalls, destroy fishing grounds, cause siltation, and cause drastic changes in the composition of the biotic community.

Geothermal energy — heat generated by natural processes from the Earth's vast subsurface storehouse of heat — includes dry steam, wet steam, and hot water deposits. Dry steam is the rarest and most preferred geothermal resource and is also the simplest and cheapest form for generating electricity. A hole is drilled into the dry steam reservoir, and the released steam is filtered to eliminate solid materials and then piped directly to a turbine. Wet steam deposits, which consist of a mixture of steam, water droplets and impurities such as salt, are more common than dry steam deposits but are also more difficult to use. The water in a wet steam deposit is superheated; its temperature is far above the boiling point of water at normal atmospheric pressure. When a wet steam deposit is drilled and brought to the surface, a fraction of the water vaporizes instantly because of the decrease in pressure. The water-steam mixture is spun in a separator to remove the steam, which is then purified and used to drive a turbine. Hot water deposits, such as those found in abundance in Iceland, are the most common type of geothermal energy. Geothermal energy does not play a major role in energy production globally, but 21 countries do realize a significant cumulative energy yield from this source. Geothermal energy is more environmentally benign than fossil fuels or nuclear power, and its associated costs are moderate. However, geothermal resources are nonrenewable on a human time scale, and there are relatively few accessible deposits. In addition, substances dissolved in the steam and water may affect air and water quality, and emptying underground deposits may cause the overlying land to become unstable.

Ocean power, a relatively untapped energy resource, includes the power of tides, waves, deep ocean currents, and on- and off-shore winds. Of these, only tidal and wave power have been used commercially. To harness tidal power, barriers or dams are built across inlets or estuaries; the barriers house a series of electric-generating turbines. Depending on the system design, electricity can be generated during the ebb or flow tide. Wave energy plants use a variety of methods for converting wave energy to electricity, including the oscillating water column (OWC). Another form of ocean power, ocean thermal energy conversion (OTEC), shows promise for power generation and desalination and aquaculture; after further development, OTEC could benefit coastal regions worldwide.

Biomass is the energy captured and stored by living organisms and thus is derived directly or indirectly from plant photosynthesis. One of the oldest and most versatile forms of energy, its primary sources are wood and wood processing residues; crop residues; animal waste products; garbage; and energy crops such as seaweed and kelp. Biomass is particularly important in LDCs; in some areas, it is exploited to such a degree that forests and land health are threatened. Biomass can also be used to produce clean-burning alcohol fuels, such as ethanol. These fuels could serve as an important alternative to gasoline, because they are both renewable and cleaner-burning, thus reducing pollution. Biogas, the mixture of gases released by decaying plant matter and animal waste, offers a further use for biomass. Methane is the chief component of biogas, which can be captured and used as a boiler fuel; it is fairly common in some countries, notably China. The advantages of biomass energy are that it is clean, readily accessible, and fairly inexpensive. However, biomass energy is a potential threat to land health when overexploited.

A direct result of human activity, solid wastes hold much potential as an energy resource. Using solid waste to produce energy (accomplished by burning refuse) is called trash conversion. Another source of energy derived from solid wastes is biogas. Biogas digesters are being installed in urban landfills with increasing regularity. Pipes drilled into landfills siphon the gas into storage tanks.

The exploitation of alternative energy sources, which may seem like a new and innovative development, is really an ancient practice. For centuries, humans have used solar radiation from the sun and heat from the Earth and have harnessed the power of winds, water, and ocean tides. Throughout the twentieth century, the widespread availability of inexpensive oil, natural gas, and coal supplies has caused a decrease in the use of alternative sources. However, declining oil reserves and increased environmental concerns associated with fossil fuels and nuclear power are spurring renewed interest in alternative energy.

Key Terms

active solar system	binding force
biogas	biomass
containment vessel	fission
geothermal energy	hydropower
isotope	nuclear energy
ocean power	ocean thermal energy conversion
passive solar system	photovoltaic cell
reactor core	solar energy
solar pond	solid waste
trash conversion	wind farm
wind power	wind turbine

Environmental Success Story

Consumers, businesses and governments around the world are finding ways to profit while simultaneously slashing their use of wood, metal, stone, plastic and other materials, reports a new study from the Worldwatch Institute.

"Groups as different as neighborhood associations and corporations are discovering that economic well-being is not necessarily linked to using vast quantities of materials," notes Gary Gardner, senior researcher at Worldwatch and co-author with Payal Sampat of *Mind Over Matter: Recasting the Role of Materials in Our Lives*. Some firms, including the Xerox Corporation and Interface Inc., a manufacturer of floor tiles, are now supplying customers with services, rather than making and selling goods. The companies lease their copiers and carpet tiles, taking the products back for recycling or remanufacture. Their materials thus circulate much longer — requiring a minimum of virgin material and generating a minimum of waste. Xerox estimates that its remanufacturing program diverted 30,000 tons from landfills in 1997 alone. Interface, which has cut landfill factory wastes by 60 percent since 1995, achieved a 20 percent jump in sales between 1995 and 1996 with virtually no increase in materials use — and posted record profits.

Source: Environmental News Network Staff. "Companies slash raw materials to up profits." *CNN interactive* (21 December 1998). Available: http://cnn.com/tech /science/9812 /21/corporate.recycle.enn/index.html. 11 January 1999.

True/False

1. Electricity generated from nuclear power plants is significantly cheaper than energy generated from fossil fuels.   T   F

2. Although they once showed great promise, photovoltaic cells are no longer thought to be a useful technology.   T   F

3. After nuclear power, wind power is the second fastest growing energy source in the world.   T   F

4. Burning is the least efficient means of converting biomass to energy.   T   F

5. Nuclear power accounts for about 20 percent of electric production.   T   F

Fill in the Blank

1. The isotope that fuels nuclear reactors is _____.

2. Passive solar systems rely on the natural forces of _____, _____, and _____.

3. The average minimum wind speed required for wind power systems is _____.

4. Three types of geothermal energy deposits are _____, _____, and _____.

5. The production of sugarcane produces a residue known as _____.

Multiple Choice

Choose the best answer.

1. All of the following are used as coolants in nuclear reactors except

   A. helium.

   B. uranium dioxide.

   C. heavy water.

   D. liquid metal.

2. A passive solar system can satisfy as much as _____ of a home's heating and cooling needs.

   A. 80 percent

   B. 90 percent

   C. 100 percent

   D. None of the above is true.

3. Which renewable source of energy produces the greatest proportion of electric-generating capacity worldwide?

   A. geothermal power

   B. solar power

   C. wind power

   D. hydropower

4. A potentially adverse environmental impact of large tidal power plants is

   A. land subsidence.

   B. reduced tidal current flow.

   C. increased tidal range.

   D. both B & C.

5. Which of the following energy sources is NOT primarily composed of biomass?

   A. agrifuel

   B. ethanol

   C. biogas

   D. crop residues

Short Answer

1. What is the result when an atom is bombarded by free neutrons?

2. What is deuterium oxide? What is its role in producing nuclear energy?

3. What is a power tower?

4. What is trash conversion?

5. List some advantages to biomass.

Thought Questions

Develop a complete answer for each of the following.

1. Describe the process used to prepare Uranium-235 for use in a nuclear reactor. Why is U-235, a relatively rare isotope, used instead of a more common form?

2. Compare the use of nuclear energy in the U.S. in the 1970s (before the OPEC oil embargo) and its use today. How does the U.S.'s use of nuclear energy compare with that of the rest of the world?

3. Explain the major arguments for and against the use of nuclear energy.

4. Describe two types of energy-producing systems that are primarily solar driven. How are they different? How are they similar? What are the advantages and disadvantages of each?

5. Why isn't wind power used as a major energy source today? Why is it less commonly used today than it was a century ago? Do you expect use of wind power to increase or decrease in the future? Why?

6. What is refuse-derived fuel? What problems have accompanied its use to generate electricity?

7. Compared to fossil fuels, are alternative energy sources better? Worse? The same? Explain your answer.

■ Related Concepts

Describe the relationship. (There may be more than one.)

BETWEEN...	AND...
active solar system	passive solar systems
hydroelectric production in MDCs	hydroelectric production in LDCs
watershed management	large-scale dams

Did You Know...?

A U.S. child consumes 30-40 times more natural resources than a child in the developing world.

Suggested Activities

1. Find a solar heating system and discover how it works.

2. Do you know where the electricity you use in your home comes from? Determine if your local utility company produces electricity from any alternative sources.

3. Use daylight whenever possible. Many tasks — reading, writing, studying — can be done by the light of a window.

UNIT **IV**

An Environmental Necessity:
Protecting Biospheric Components

Chapter 14
Air Resources

 Chapter Outline

I. Describing Air Resources: Physical and Biological Characteristics

  A. What Is the Atmosphere?

  B. How Does the Atmosphere Help to Maintain the Earth's Climate?

   1. Albedo

   2. Emissivity

   3. Greenhouse Effect

   4. Global Climate Change

II. Describing Air Resources: Social Characteristics

  A. What Is Air Pollution?

   1. Primary Pollutants

   2. Secondary Pollutants

  B. What Is Climate Change?

   1. Trends in Greenhouse Gases and Average Global Temperature

   2. Forecasts of Expected Climate Change

   3. Potential Effects of Climate Change

   4. Scientific Evidence Supporting Climate Change

   5. Efforts to Combat Climate Change

  C. What Is Stratospheric Ozone Depletion?

   1. Ozone-Destroying Chemicals

   2. Effects of Stratospheric Ozone Depletion

   3. Efforts to Combat Stratospheric Ozone Depletion

  D. What Is Acid Precipitation?

   1. Formation and Transport of Acid Precipitation

   2. Effects of Acid Precipitation

   3. Efforts to Combat Acid Precipitation

    E. What Is Smog?

    F. What Are Air Toxics?

    G. What Is Indoor Air Pollution?

    H. What Factors Affect Air Pollution Levels?

        1. Weather

        2. Topography

        3. Temperature Inversions

III. Managing Air Resources

    A. What Is the Clean Air Act?

        1. Setting Emission Standards

        2. Meeting Emission Standards

        3. Establishing Air Quality Standards

        4. Monitoring Air Quality Standards

        5. Improving Air Quality

        6. Air Toxics

        7. Acid Precipitation

    B. What Problems Surround Air Resources Worldwide?

Learning Objectives

After learning the material in Chapter 14, you should be able to:

1. Describe the composition of the atmosphere and discuss how it is affected by living systems.

2. Distinguish between primary and secondary air pollutants and describe their sources.

3. Identify and briefly discuss six primary air pollutants and two major secondary pollutants.

4. Identify and describe the characteristics that affect air pollution levels.

5. Describe the effects of air pollutants on environmental and human health.

6. Summarize the methods used to control air pollution.

Did You Know...?

The number of cars in the world is growing about three times faster than the number of humans.

 ## Key Concepts

Read this summary of Chapter 14 and identify the important concepts discussed in the chapter.

The atmosphere consists of layers of gases that surround the Earth. Beginning at the Earth's surface, these layers are the troposphere, stratosphere, mesosphere, and thermosphere. The troposphere, which contains the gases that support life, is also the layer in which weather occurs. Weather is the day-to-day patterns of precipitation, temperature, wind (direction and speed), barometric pressure, and humidity. The long-term weather pattern of a region is known as its climate. Both regional climates and the global climate are heavily influenced by the atmosphere. Atmospheric gases such as water vapor and carbon dioxide allow some incoming solar radiation to pass through the atmosphere, but also trap heat that would otherwise be reradiated back toward space, a phenomenon known as the greenhouse effect. Greenhouse gases are essential to maintaining the conditions necessary for life on Earth, but increased levels of carbon dioxide, methane, chlorofluorocarbons, and nitrous oxide can lead to global warming.

An air pollutant is defined as any substance present in or released to the atmosphere that adversely affects human health or the environment. Those emitted directly into the atmosphere are known as primary pollutants. Secondary pollutants are formed when primary pollutants react with other primary pollutants or with compounds present in the atmosphere, such as water vapor. Of the thousands of primary pollutants, six are of special concern because they are emitted in particularly large quantities: carbon dioxide, carbon monoxide, sulfur oxides, nitrogen oxides, hydrocarbons, and particulates. In the 1970s, efforts at cleaning up air pollution in the United States focused on primary pollutants. Now, however, increased attention is being given to six issues or problems — many of them related to the formation of secondary pollutants — that are particularly troublesome: global climate change, stratospheric ozone depletion, acid precipitation, smog, airborne toxins, and indoor air pollution.

Increased atmospheric concentrations of greenhouse gases, especially carbon dioxide, are expected to cause a general warming, thus changing the global climate. Indeed, over the last 100 years, as atmospheric carbon dioxide increased about 30 percent, the global temperature increased by about 1°F. If present trends continue, atmospheric CO_2 concentrations will double from preindustrial levels by around 2075. A doubling of atmospheric CO_2 would lead to a 9°F increase in average world temperature within 20 to 30 years. The most likely effects of global warming are changes in rainfall patterns, changes in growing seasons, changes in arable land, and rising sea levels. Some studies have also shown that climate change is already causing glaciers to melt and shrink in size, affecting the influx of glacial melt into surface waters. Other studies have also demonstrated a probable effect on living organisms; some species' range appears to be shrinking while others' range may be expanding.

Ozone in the stratosphere plays a critical role in protecting the Earth from harmful ultraviolet radiation. That protective layer, however, is threatened by human made substances known as chlorofluorocarbons (CFCs). The breakdown products of CFCs, chlorine and fluorine, along with nitrogen oxides and bromines, can destroy ozone. The stratospheric ozone layer is extremely thin, so even small decreases are expected to cause significant increases in cancer rates among humans and decreases in marine productivity. Under the terms of the Montreal Protocol, the nations of the world have begun to phase out the use of CFCs and other ozone-destroying chemicals. Unfortunately, progress in restoring the ozone layer has been slowed by a growing black-market trade in CFCs.

Acid precipitation is rain, snow, fog, or mist that contains enough sulfuric acid, nitric acid, or their precursors (sulfur and nitrous oxides) to raise the acidity of the precipitation above normal. Its adverse effects are many and widespread: Acid precipitation can damage lakes, forests, and crops; threaten human health; and corrode marble, limestone, sandstone and bronze, defacing or destroying statues, monuments, gravestones, and buildings. Fish and amphibians such as salamanders and frogs may be especially vulnerable to acid precipitation, particularly during an acid surge — a period of short, intense acid deposition in lakes and streams, which may occur after a heavy rain or a snowmelt. The damage caused by acid precipitation is widespread, both in the United States and globally (in such regions as Northern and Eastern Europe and China). Although U.S. laws have lowered emissions of sulfur dioxide, emissions of nitrous oxides appear to be increasing (likely from the increased number of vehicles on U.S. roads), offsetting substantial gains that may have been made in the actual occurrence of acid precipitation. Moreover, it's likely that the ill effects of acid precipitation are compounded by other atmospheric problems (such as photochemical smog and climate change) and therefore cannot be remedied without measures to combat those contributing factors.

There are two main types of smog, industrial and photochemical. Industrial smog is essentially smoke pollution; it consists chiefly of sulfur oxides and particulates and is emitted to the air from industrial and manufacturing facilities. In the United States, it is particularly problematic in the industrial Northeast. Photochemical, or brown, smog is formed when hydrocarbons react with nitrogen oxides and oxygen to form chemicals such as ozone and peroxyacetyl nitrate (PAN). Photochemical smog is a serious problem in urban areas worldwide; in the United States, Los Angeles is notorious for poor air quality related to smog. The primary component of photochemical smog is ozone that occurs at ground level. Ozone is toxic to living organisms; it can adversely affect human health in many ways and can also damage crops and forests.

Hazardous air pollutants, commonly called air toxics, are a varied group of chemical substances, numbering in the hundreds, known to cause or suspected of causing cancer or other serious human health effects or ecosystem damage. Relatively few of these pollutants have been studied for their effect on human health or the environment.

Indoor air pollutants — pollutants found in homes, schools, and businesses — pose a serious human health problem, a problem that grows as more people spend more time indoors and as construction methods have resulted in tighter, more energy-efficient homes (which may result in a lack of fresh air entering a building). Major indoor air pollutants include formaldehyde, radon 222, tobacco smoke, asbestos, combustion products from gas stoves and poorly vented furnaces, chemicals used in building and consumer products, and disease-causing organisms or spores. In the United States, the dramatic increase in asthma cases over the past several decades is believed to be linked to indoor air pollution.

Those most at risk from air pollution are the very old, the very young, and persons already suffering from respiratory or circulatory disease. Generally speaking, air pollution is worse in urban and suburban areas (because of the concentration of industry and the large number of vehicles) and can be especially problematic in improperly ventilated buildings.

Pollution levels are affected by weather and topography. Secondary pollutants can be formed in the presence of sunlight or when primary pollutants react with moisture in clouds or highly reactive chemicals present in the atmosphere. Topography, or the lay of the land, can also act to concentrate pollution: Valleys act as sinks for pollutants and mountains act as barriers, hindering the flow of air currents that would otherwise carry away and disperse pollutants.

In the United States, air pollution is regulated chiefly by the Clean Air Act, first passed in 1963 and amended in 1970, 1977, and 1990. The Clean Air Act directed the EPA to develop national ambient air quality standards (NAAQS), emission standards that specify the quantities of air pollutants that can be emitted by specific sources. Its later amendments contain provisions concerning smog, airborne toxins, and acid precipitation. In contrast, many other countries, especially in the developing world, have few regulations in place to limit or control air pollution.

 Key Terms

acid precipitation	acid surge
aerosols	air pollutant
air quality index	albedo
asbestos	atmosphere
climate	climate change

emission standards

greenhouse effect

indoor air pollution

mesosphere

ozone

primary pollutant

rain shadow effect

stratosphere

thermosphere

weather

emissivity

hazardous air pollutants

industrial smog

national ambient air quality standards

photochemical smog

radon 222

secondary pollutant

stratosphere ozone depletion

troposphere

Environmental Success Story

UCLA is helping to combat Los Angeles' smog problem through the Energy Systems Facility. Designed to heat and cool the campus, this independent source of electric energy conserves natural resources in four ways: by increasing energy efficiency, by utilizing landfill gas, by recycling water, and by saving energy. Chilled water units have been replaced by a cogeneration system, and overall campus atmosphere emissions have decreased by 34 percent. In addition, UCLA has reduced its use of natural gas by one-third and saves 70 million gallons of water annually through its use of recovered water.

Source: Energy Systems Facility. "Seventh Annual National Awards for Environmental Sustainability Winners: Atmosphere and Climate." *Renew America* (22 January 1997). Available: http://solstice.crest.org/environment/renew_america/winners.html. 18 February 1999.

True/False

1. Air is composed primarily of oxygen.   T   F

2. Areas with a low albedo, such as polar ice caps and deserts, reflect more energy than they absorb.   T   F

3. Average global temperature is about 59°F.   T   F

4. Secondary pollutants are those that are released to the atmosphere in relatively small quantities.   T   F

5. The primary component of photochemical smog is ground level ozone.   T   F

Fill in the Blank

1. Infrared ray-absorbing gases in the atmosphere are often referred to as _____.

2. Short, intense periods of acid deposition in lakes and streams are called _____.

3. Manufacturing facilities emit sulfur oxides and particulates known as _____.

4. The phenomenon that causes one side of a mountain to be damaged by pollution while the other side remains unharmed is known as _____.

5. The Earth's atmosphere consist of the _____, _____, _____, and _____.

Multiple Choice

Choose the best answer.

1. The atmospheric level in which weather occurs is the
   A. mesosphere.
   B. stratosphere.
   C. thermosphere.
   D. troposphere.

2. The capacity of a surface to radiate heat is its
   A. albedo.
   B. emissivity.
   C. radiational effect.
   D. greenhouse effect.

3. Which of the following is NOT true of acid precipitation?
   A. It occurs in the form of rain, snow, fog, mist or dust.
   B. It has a pH of 5.5 or lower.
   C. Its low pH results primarily from sulfuric and carbonic acid.
   D. None of the above is true.

4. All of the following are major indoor air pollutants except

   A. carbon dioxide.

   B. formaldehyde.

   C. asbestos.

   D. radon 222.

5. Ozone is an air pollutant in which of the following places?

   A. indoors

   B. troposphere

   C. stratosphere

   D. All of the above are true.

Short Answer

1. List the components of clean, dry air.
2. What are the six primary air pollutants?
3. What are the factors that affect air pollution levels?
4. What are the causes of temperature inversions?
5. What is the largest unregulated source of air pollution in the U.S.?

Thought Questions

Develop a complete answer for each of the following.

1. Explain the importance of the greenhouse effect in maintaining life on Earth. List the causes of increased greenhouse gases and describe the effects that may result.
2. What evidence exists that global climate change is already occurring? What are the arguments and uncertainties that point to the contrary? Discuss the general agreements (the "greenhouse knowns") among the scientific community.
3. Discuss the causes and effects of acid precipitation. Why is it such a widespread problem? What are the political problems associated with regulating it?
4. Explain the "split personality" of ozone. Include in your answer a discussion of the effects of ozone on human health.
5. What are CFCs and how are they used? What is their impact on the environment?
6. Discuss the effects of indoor air pollution. What factors make this a difficult type of pollution to control?

7. Describe the history of the air quality legislation in the U.S. Has it been effective? Why or why not?

▢▨■ Related Concepts

Describe the relationship. (There may be more than one.)

BETWEEN...	AND...
greenhouse gases	global climate change
soil composition	acid precipitation
topography & weather	air pollution
airborne toxics	water pollution
standard of living	air quality in the U.S.

Did You Know...?

As a nation, Americans drive the distance to Pluto and back, every single day.

 ## Suggested Activities

1. Measure radon levels in your home. Refer to EPA guidelines in deciding whether and how quickly to take action based on your test results.

2. Stop smoking and discourage smoking in your home.

3. If you have your car's air conditioner serviced, choose a car repair shop that uses CFC recycling equipment.

Chapter 15
Water Resources

 Chapter Outline

I. **Describing Water Resources: Biological Characteristics**

   A. How Does Water Support Life?

   1. Water is the largest constituent of living organisms and provides habitat for a great diversity of life.

   2. The human body is composed primarily of water (about 65 percent).

   3. Water makes up 83 percent of human blood.

II. **Describing Water Resources: Physical Characteristics**

   A. How Much Clean Freshwater Is There?

   B. How Is Water Classified?

   1. Freshwater

      a. Surface waters

         i. Standing water habitats (oligotrophic and eutrophic lakes)

         ii. Running water habitats (streams and rivers)

      b. Groundwater

         i. Aquifers (confined and unconfined)

   2. Marine Waters

      a. Euphotic zone

      b. Neritic zone

      c. Pelagic zone

      d. Abyssal zone

III. **Describing Water Resources: Social Characteristics**

   A. How Do We Use Water?

   1. To drink, bathe, dispose of wastes, irrigate crops, support industry, and generate power

   B. How Do Water Use Patterns Vary Worldwide?

1. In the U.S., agricultural and industrial use dropped by two percent between 1990 and 1995, but domestic use increased by three percent.

2. Disease prevention through an adequate supply of clean water is the key to improving health standards in LDCs.

C. What Kinds of Water Pollution Are There?

1. Organic Wastes

2. Disease-Causing Wastes

3. Plant Nutrients

4. Sediments

5. Toxic And Hazardous Substances

6. Persistent Substances

7. Radioactive Substances

8. Heat

IV. Managing Water Resources

A. How Have Humans Managed Water Resources in the Past?

1. Two Axioms Concerning Water

    a. Draw water from upstream and deposit wastewater downstream.

    b. If the habitat is wet, make it dry; if it is dry, make it wet.

B. What Problems Surround the Use of Water Worldwide?

1. Water Stress

2. Water Scarcity

C. How Have Water Resources Been Managed in the United States?

1. New York

2. Los Angeles

3. New Orleans

D. What Is the Safe Drinking Water Act?

E. What Is the Clean Water Act?

F. How Are Drinking Water Supplies Treated?

G. How Is Wastewater Treated?

1. Septic Tank

2. Sewer

                a. Combined sewer system

                b. Separate sewer system

                c. Multistage treatment process (physical, biological, chemical)

        3. Management Issues Associated With Wastewater Treatment

    H. How Is Groundwater Managed?

        1. Overdraft

        2. Water Mining

        3. Land Subsidence

    I. How Are Fresh Surface Waters Managed?

        1. Measures To Protect Fresh Surface Waters

    J. How Are Marine Waters Managed?

        1. Measures To Protect Marine Waters

Learning Objectives

After learning the material in Chapter 15, you should be able to:

1. List the properties of water that enable it to support life.

2. Identify how water resources are classified and briefly describe each classification.

3. Describe worldwide water consumption patterns.

4. Identify eight broad categories of water pollution.

5. Discuss major legislation concerning water resources in the United States.

6. Describe how both drinking water supplies and wastewater are treated.

7. List some of the current threats to groundwater, lakes, rivers, and marine waters.

> **Did You Know...?**
>
> Water-saving devices can shrink your water use by 30,000 gallons per year.

Key Concepts

Read this summary of Chapter 15 and identify the important concepts discussed in the chapter.

Most of the Earth's water (97 percent) is salty. Three-quarters of the remaining freshwater is locked in polar ice caps and glaciers, and one-quarter is found underground as groundwater. Only 0.5 percent of all water in the world is found in lakes, rivers, streams, and the atmosphere.

Water is classified as either salt (marine) water or freshwater, depending on its salt content. Freshwater is found on land in two basic forms: surface water and groundwater. Surface waters include all bodies of water that are usually recharged by runoff from precipitation. Surface water ecosystems include both standing water habitats, which are relatively closed ecosystems with well-defined boundaries, and running water habitats, which are continuously moving currents of water.

An example of standing water habitats, lakes are categorized according to the amount of dissolved nutrients they contain. Oligotrophic lakes are cold, blue, and deep. They contain low amounts dissolved solids, nutrients, and phytoplankton. Eutrophic lakes are warmer and more turbid and have a lower oxygen content. They are far more productive.

Running water habitats include streams and rivers. The speed and temperature of the current affect the kinds of organisms that live in a habitat. Fast-moving rivers and streams usually contain two kinds of habitats: riffles and pools. Riffles, with a high oxygen content, tend to house the producers of biomass, and pools tend to contain consumers and decomposers.

The United States contains about 20 times more groundwater than surface water. Groundwater percolates downward through the soil after precipitation or from surface water and is stored in an aquifer. Groundwater is the major source of drinking water in about two-thirds of the states. In addition, it helps to maintain water levels and the productivity of streams, lakes, rivers, wetlands, bays, and estuaries.

Oceans support over half of the world's biomass and are one huge living system. Currents move water around the continents and across the open seas, continually circulating nutrients washed in from the land. The interaction of the oceans and the atmosphere affects heat distribution, weather patterns, and concentrations of atmospheric gases throughout the world. Because they are so expansive and deep, the oceans moderate climate and provide a sink for dissolved solids and gases. Beneath the ocean floor lies a rich storehouse of minerals, petroleum, and natural gas. Hundreds of millions of people live near the ocean shores and on coastal plains. Even greater numbers depend upon ocean fisheries, energy, and minerals for their sustenance.

Water uses may be consumptive or nonconsumptive. Nonconsumptive uses remove water, use it, and return it to its original source. Consumptive uses remove water from one place in the hydrological cycle and return it to another. We use water for many purposes, but globally, agriculture is the single greatest drain on water supplies.

Water pollutants can be divided into eight general categories: organic wastes, disease-causing wastes, plant nutrients, toxic and hazardous substances, persistent substances, sediments, radioactive substances, and heat. Some of these categories overlap. One source may be responsible for more than one type of pollutant, or one pollutant may fit into more than one category. Pollutants can also act together synergistically. Many water systems are assaulted by pollutants from all eight categories.

Worldwide, water resources are the source of hundreds of environmental and economic problems. Water stress (the episodic lack of renewable freshwater availability) and water scarcity (the chronic lack of renewable freshwater availability) are at the heart of these problems. Regions in the Middle East and northern and eastern Africa face an especially difficult situation, as rapid population growth further strains water resources that are already severely limited by an arid climate and shared access to rivers.

Legislation, or the lack of it, determines how water is managed in the United States. The Safe Drinking Water Act of 1974 set national drinking water standards, called maximum contaminant levels (MCLs), for pollutants that might adversely affect human health. It also established standards to protect groundwater from hazardous wastes injected into the soil. The 1986 reauthorization of the act instructed the EPA to monitor drinking water for unregulated contaminants and to inform public water suppliers concerning which substances to look for. The 1996 reauthorization built upon the 1986 regulations, requiring the EPA to identify and monitor five new contaminants every five years and to provide residents with annual "Consumer Confidence Reports" that detail the status of the water they drink. The 1972 Federal Water Pollution Control Act, commonly known as the Clean Water Act, divided pollutants into three classes: toxic, conventional, and unconventional. The act stipulated that industries must use the best available technology (BAT) to treat toxic wastes before releasing them into natural waters and the best conventional technology (BCT) to treat conventional pollutants, such as municipal wastes. Unconventional pollutants must meet BAT standards. The 1987 reauthorization of the Clean Water Act granted the Army Corps of Engineers the authority to regulate the draining and filling of wetlands, though that authority was restricted when a federal judge overturned an amendment commonly known as the "Tulloch Rule."

After water is withdrawn from a lake, river or aquifer, it is treated before being distributed. In areas that have suitable soils, sewage and wastewater from each home is usually discharged into a septic system consisting of an underground tank made of concrete and a drain field. Wastewater in urban areas must be collected in underground sewers and treated in sewage treatment plants. These plants are designed to make wastewater safe for discharge into streams or rivers or to make it acceptable for reuse. Most sewage plants employ a multistage process to reduce wastewater to an acceptable effluent.

Fresh and marine waters worldwide are besieged by numerous threats. Many groundwater aquifers experience overdraft, the withdrawal of water faster than the aquifer can be recharged. Overdrafts can lead to land subsidence. When groundwater becomes contaminated, it is a costly, slow, and sometimes impossible task to remove pollutants. Toxic pollution and accelerated eutrophication are serious threats to the United States's lakes. Threats to rivers include dams, diversions, channelization, and pollution. Abuse of the ocean ecosystem can come from pollution and overuse. All U.S. coastal ecosystems are under severe stress. Until recently, little thought was given to effectively managing and preserving marine environments.

Key Terms

abyssal zone	adsorption
aquifer	biological oxygen demand (BOD)
combined sewer system	confined aquifer
desalination	epilimnion
euphotic zone	eutrophic
groundwater	hypolimnion
land subsidence	limnetic zone
littoral zone	neritic zone
oligotrophic	overdraft
pelagic zone	profundal zone
running water habitat	runoff
separate sewer system	septic tank
standing water habitat	surface waters
thermal stratification	thermocline
unconfined aquifer	water mining
water scarcity	water stress
water table	watershed

Environmental Success Story

In the early 1990s, the town of Gilbert, Arizona, was running out of water. Instead of attempting to develop new resources, local officials saved money and protected the environment by using their existing supply more efficiently. To meet the goal of reclaiming 100 percent of their water, the town built eight ponds that recharge groundwater on a 38-acre site. In the ensuing three years, facility managers noticed that local bird watchers were frequenting the site. When it came time to expand the recharge basin, the city developed the basin into an urban wildlife habitat, hoping to attract additional species indigenous to a riparian environment. They constructed educational trail signs and homes for bats, hummingbirds, and other animals. Currently, numerous species have settled in the area, enabling local school children and college students to study a wide array of organisms. Most importantly, the town of Gilbert has proven that with a little creativity, municipal infrastructure can be designed to enhance both the community and the environment.

Source: Town of Gilbert. "Eighth Annual National Awards for Environmental Sustainability Winners: Urban Wildlife Habitat and Groundwater Recharge Facility." *Renew America* (22 April 1998). Available: http://solstice.crest.org/sustainable/renew_america/winner98.html. 24 March 1999.

True or False

1. Only about three percent of the Earth's water is freshwater.   T   F
2. Domestic water use decreased in the U.S. between 1990 and 1995.   T   F
3. Aquifers supply drinking water for half of the U.S. population.   T   F
4. Secondary treatment of wastewater is usually about 85 percent effective.   T   F
5. Groundwater withdrawal in the U.S. has tripled since the 1950s.   T   F

Fill in the Blank

1. Cold, deep lakes with a high oxygen content are called _____ lakes.
2. The layers of thermal stratification (Figure 15-3) are the _____, _____, and _____.
3. The ocean zones are _____, _____, _____, _____, and _____.
4. The amount of oxygen needed to decompose the organic matter in water is called _____.
5. By weight, _____ are the most abundant water pollution.

Multiple Choice

Choose the best answer.

1. Most of the Earth's freshwater is contained in

    A. lakes, rivers, and streams.

    B. the atmosphere.

    C. aquifers.

    D. None of the above is true.

2. The shallow area near the shore of a lake where rooted plants grow is the

    A. profundal zone.

    B. limnetic zone.

    C. littoral zone.

    D. pelagic zone.

3. In the oceans, phytoplankton grow most abundantly in the

    A. pelagic zone.

    B. littoral zone.

    C. euphotic zone.

    D. neritic zone.

4. The test used to determine if water is likely to be contaminated with disease-causing organisms looks for the presence of

    A. coliforms.

    B. nitrosamines.

    C. giardiasis.

    D. algal blooms.

5. A feature of tertiary treatment of wastewater is

    A. activated carbon adsorption.

    B. activated sludge.

    C. removal of undissolved solids.

    D. None of the above is true.

Short Answer

1. What is a watershed?
2. List the defining characteristics of eutrophic lakes.
3. What are unconfined aquifers and confined aquifers?
4. Define effluent.
5. What chemicals are commonly added to drinking water supplies and why?

Thought Questions

Develop a complete answer for each of the following.

1. Discuss the human health and environmental problems associated with water consumption worldwide.
2. List and briefly describe the effects of the eight types of water pollution.
3. Explain the U.S. legislation designed to protect water quality.
4. Describe the process of wastewater treatment. Include in your answer a comparison of combined and separate sewer systems. What occurs at each stage (primary, secondary, tertiary) of treatment?
5. What is the importance of groundwater resources in the U.S.? What are the major threats facing these resources?

▢▨■ Related Concepts

Describe the relationship. (There may be more than one.)

BETWEEN...	AND...
seasons	thermal stratification
riffles and pools	running water habitats
organic waste	eutrophication
irrigation	channelization
overdraft	land subsidence
water stress	water scarcity

Did You Know...?

Worldwide, 70 percent of all water used by humans is for irrigation.

 Suggested Activities

1. Contact your water company to ask for the most recent analysis of the compounds and chemicals found in your drinking water, or obtain your own water analysis. Compare your results with the national limits set by the EPA.

2. Does your county or town have its own water quality standards? Find out what they are and how they are enforced.

3. Investigate the effect of oil spills on aquatic habitats. Examples of recent spills include the *Exxon Valdez* spill in Prince William Sound, Alaska, and the Persian Gulf spills that occurred during the 1991 expulsion of Iraqi troops from Kuwait.

4. Look for and eliminate leaks around your home.

Chapter 16
Soil Resources

 Chapter Outline

I. **Describing Soil Resources: Biological Characteristics**

    A. What Is Soil?

        1. Topmost Layer Of The Earth's Surface

        2. Soil Fertility

        3. Soil Productivity

    B. How Do Organisms Maintain Soil Fertility?

II. **Describing Soil Resources: Physical Characteristics**

    A. What Is Soil Texture?

        1. Mineral content determines texture, or feel, of soil.

    B. What Is Soil Structure?

        1. Arrangement of soil particles

    C. How Is Soil Formed?

        1. Parent Material

        2. Climate

        3. Topography

        4. Living Organisms

        5. Time

    D. What Is a Soil Profile?

        1. Soil horizons — horizontal layers, each with distinct characteristics

III. **Describing Soil Resources: Social Characteristics**

    A. How Is Land Used?

    B. How Do Land Uses Affect the Soil?

        1. Suburbanization

        2. Erosion

            a. Soil loss tolerance level (T-value)

IV. **Managing Soil Resources**

    A. How Have Agricultural Lands Been Managed in the United States?

    B. What Was the Dust Bowl?

    C. What Environmental Problems Are Associated with Conventional Agriculture?

        1. Soil Degradation

        2. Soil Erosion

        3. Reliance On Fossil Fuels

        4. Reliance On Synthetic Chemicals

        5. Groundwater Depletion

        6. Poor Irrigation Practices

        7. Loss Of Genetic Diversity

    D. What Was the Farm Crisis of the 1980s?

    E. What Is the Food Security Act?

    F. What Is the Federal Agriculture Improvement and Reform Act?

    G. What Is Sustainable Agriculture?

        1. Measures To Prevent Excessive Soil Erosion

        2. Measures To Enhance Soil Fertility

        3. Organic Farming

        4. Polyculture

Learning Objectives

After learning the material in Chapter 16, you should be able to:

1. Describe the major components of soil and explain how soil is formed.

2. Describe a typical soil profile.

3. List the major uses for land and describe how land uses affect agriculture and the environment.

4. Compare conventional and sustainable agriculture practices.

5. Discuss methods to prevent soil erosion and maintain soil fertility.

Did You Know...?

A standard shower head uses five to seven gallons of water per minute, so even a five minute shower can consume thirty-five gallons!

 Key Concepts

Read this summary of Chapter 16 and identify the important concepts discussed in the chapter.

Soil, the topmost layer of the Earth's surface, is an ecosystem composed of abiotic and biotic components, such as inorganic chemicals, air, water, decaying organic material, and living organisms. As such, it is subject to the dynamics that operate in all ecosystems. Humus, which consists of partially decomposed organic matter, helps to retain water and maintain a high nutrient content, thus enabling soil to remain fertile. Soil fertility refers to the soil's mineral and organic content, while soil productivity refers to its ability to sustain life, especially vegetation. Every teaspoon of soil contains billions of beneficial organisms that help maintain soil fertility.

Soil varies in texture, structure, and type. Soil texture is the way the soil feels. Soil structure, or tilth, refers to the way soil particles clump together to form larger clumps. Scientists typically recognize ten major soil orders and an estimated 100,000 soil types. The soil types have been formed through the interaction of five factors: parent material, climate, topography, living organisms, and time. As soils develop, they form distinct horizontal layers called soil horizons. A vertical series of soil horizons in a particular location is a soil profile.

The way in which a particular piece of land is used is known as its land use. Depending on the soil type and the terrain, there may be many potential uses for a specific parcel of land. The use to which a piece of land is put can have immediate and long-lasting effects on soil resources. Worldwide, expanding urbanization poses a long-term threat to soil resources. An even greater threat to soil fertility and conservation is erosion. Erosion by wind and water is a natural process, and when undisturbed, soil is usually replaced faster than it erodes. Soil disturbed by human activities, however, can be eroded faster than it is replaced. Because soil takes so long to form, eroded soil is — in terms of the human life span — irreplaceable.

In the United States, no land use has been more important historically than agriculture. When the Europeans first arrived on the North American continent, they cleared seemingly inexhaustible forests for farmlands. As land became increasingly scarce and the soils in some areas became less productive, settlers began to move westward. The rich grassland soils of the Great Plains produced large yields for years, but soil fertility gradually declined. In 1931, the combination of a severe drought and intense cultivation of unsuitable crops resulted in the Dust Bowl, a period of extensive wind erosion.

In the 1950s and 1960s, agricultural production increased through the use of high inputs of chemicals, large machinery, and hybrid strains of crops. But high-input, or conventional, agriculture can cause serious problems, including soil degradation, soil erosion, reliance on fossil fuels, reliance on agrichemicals, groundwater depletion, overirrigation, and loss of genetic diversity. Agricultural corporations, the U.S. Department of Agriculture, and government programs largely have supported the trend toward highly specialized, energy-intensive, and environmentally questionable farming practices.

In part, the Farm Crisis of the 1980s was a result of these farming practices. Coupled with economic recession and high interest rates, the soaring costs of energy-intensive agriculture forced farmers to borrow heavily. Those who could not repay their loans faced the difficult decision of selling their farms or watching as the bank foreclosed on their land. The rise of industrial agriculture throughout the 1990s continues to stress the livelihood of America's small farmers.

The Food Security Act of 1985, also called the Farm Bill, has been hailed as the cornerstone of the most progressive agricultural policy worldwide. Its provisions, especially the swampbuster, sodbuster and conservation reserve program, encourage farmers to remove highly-erodible or marginal land from production. The 1996 reauthorization of the Farm Bill aims to eliminate selected government subsidies while still protecting wetlands and keeping prime farmland in agriculture. The Bill's major provision is the Agriculture Market Transition Program, commonly called "Freedom to Farm," which creates seven-year Production Flexibility Contracts that grant farmers fixed but declining annual payments, regardless of market prices. The premise of Freedom to Farm is that the market should determine which crops are planted, not the federal government.

The search for effective methods to protect the soil and restore its fertility is resulting in the development of various systems of alternative or sustainable agriculture. Sustainable agriculture involves techniques such as growing a variety of crops, rotating crops, using organic fertilizers, allowing croplands to lie fallow periodically, planting cover crops, and encouraging the natural enemies of crop pests. Alternative methods of tilling help prevent soil erosion.

Of all forms of agriculture, organic farms most closely resemble natural systems. They are characterized by diversity. Soil texture and fertility are enhanced by organic soil amendments. Weeds are controlled by mulching and cultivating techniques. Insect pests are controlled through a variety of methods collectively known as integrated pest management. Biological pest control involves the introduction of insects, birds, animals, or microbes that attack pests. Crop rotation prevents the depletion of soil nutrients and promotes higher yields. A new agricultural approach called perennial polyculture, in which a number of perennial crops are established in an area, also renders agriculture less environmentally harmful.

Key Terms

contour plowing

crop rotation

humus

land use

macroclimate

no-till planting

ridge tilling

soil fertility

soil loss tolerance level (T-value)

soil profile

subsoil

tilth

trickle drip irrigation

zone of leaching

contour terracing

green manure

integrated pest management (IPM)

low-till planting

microclimate

perennial polyculture

salinization

soil horizon

soil productivity

strip cropping

sustainable agriculture

topsoil

windbreak

Environmental Success Story

Before 1996, Florida's pest management model for public schools was "see a bug, spray a bug" — a use of chemicals that was proving harmful to teachers and students. In response to this problem, the Legal Environmental Assistance Foundation created the Community Action to Manage Pesticide Use in Schools (CAMPUS) program, which seeks to teach schools how to reduce their use of chemical pesticides. Students learn a variety of techniques for controlling pests without chemicals, including using natural predators, such as ladybugs; sealing cracks in walls so that pests can not enter buildings; and applying nontoxic pesticide alternatives, such as soaps and oils. To date, the program, which offers workshops and guides, has reached over 300 schools in Florida.

Source: Legal Environmental Assistance Foundation, Inc. "Eighth Annual National Awards for Environmental Sustainability Winners: Pollution Prevention." *Renew America* (22 April 1998). Available: http://solstice.crest.org/sustainable/renew_america/winner98.html. 24 March 1999.

True or False

1. Soil fertility refers to the ability of a soil to sustain life (such as vegetation). T F

2. Loams are the best soil types for growing most crops. T F

3. In the days of the Dust Bowl, soil was being depleted at the fastest rate in U.S. history. T F

4. Water accounts for two-thirds of the erosion on U.S. farmland. T F

5. It takes an average of three to five years to convert a farm from conventional techniques to organic practices. T F

Fill in the Blank

1. The mineral and organic content of soil determines its _____.

2. The arrangement of soil particles into lumps or crumbs is known as _____.

3. The soil horizon that contains humus, living organisms, and some minerals is called the _____.

4. Raw mineral material from which soil is formed is called _____.

5. The _____ describes the amount of soil that a given area can lose through erosion without a loss of fertility.

Multiple Choice

Choose the best answer.

1. Silt particles
   A. are less than 0.00008 inches in diameter.
   B. feel like flour when rubbed between the fingers.
   C. are sticky when wet.
   D. All of the above are true.

2. The soil layer in which dissolved minerals accumulate is the

   A. A horizon.

   B. B horizon.

   C. E horizon.

   D. O horizon.

3. The zone of leaching is also called the

   A. A horizon.

   B. B horizon.

   C. C horizon.

   D. E horizon.

4. The first European settlers in North America used farming techniques that included

   A. continual cropping.

   B. crop rotation.

   C. irrigation.

   D. None of the above is true.

5. A technique designed to reduce erosion on sloped farmland is

   A. contour plowing.

   B. trickle drip irrigation.

   C. crop rotation.

   D. ridge tilling.

Short Answer

1. Describe the abiotic composition of soil.
2. List and describe the three soil categories based on texture.
3. What are the five interacting factors that form soil?
4. What are the two major causes of agricultural soil erosion?
5. What is a perennial polyculture?

Thought Questions

Develop a complete answer for each of the following.

1. Describe the differences between soil fertility and soil productivity. How are they maintained by organisms?

2. Which land uses are the biggest threats to soil fertility and conservation in the United States? How can land use planning help?

3. Discuss the factors and events that led to the Dust Bowl.

4. What changes occurred in American agriculture during the 1950s and 1960s? Discuss the continuing effects of these changes on the environment.

5. What is sustainable agriculture? Where and how is it being practiced today, and with what results?

6. Discuss how the U.S. government has responded to the Farm Crisis.

▢▬■ Related Concepts

Describe the relationship. (There may be more than one.)

BETWEEN...	AND...
Dust Bowl	soil conservation
soil erosion	loss of wildlife habitat
irrigation	groundwater depletion
Food Security Act of 1985	1996 Farm Bill

Did You Know...?

Americans receive almost four million tons of junk mail every year. Most of it winds up in landfills.

 Suggested Activities

1. Find out what kinds of crops are grown in your part of the country. Locate the Soil Conservation Service in your area and find out what kinds of conservation techniques are recommended.

2. If you have a vegetable garden, try using biological pest controls and other alternatives to pesticides.

3. Compare the appearance and taste of organically grown fruits and vegetables with those that are conventionally grown.

Chapter 17
Biological Resources

 Chapter Outline

I. **Describing Biological Resources: Biological Characteristics**

   A. What Are Biological Resources?

      1. Biological Diversity (Biodiversity)

         a. Species Diversity

         b. Ecosystem Diversity

         c. Genetic Diversity

   B. What Is Extinction?

      1. The elimination from the biosphere of all individuals of a group of organisms of any taxonomic rank.

   C. What Is the Relationship between Extinction and Biological Diversity?

      1. Genetic Erosion

      2. Minimum Viable Population Size

II. **Describing Biological Resources: Physical Characteristics**

   A. Which Habitats Have the Greatest Number of Species?

      1. Endemic Species

III. **Describing Biological Resources: Social Characteristics**

   A. How Do Beliefs and Attitudes Affect the Management of Biological Resources?

   B. How Do Human Activities Adversely Affect Biological Resources?

      1. Habitat Loss

      2. Overharvesting and Illegal Trade

      3. Selective Breeding

      4. Introduced Species

      5. Pollution

   C. Why Should We Preserve Biological Resources and Biological Diversity?

      1. Ecosystem Services

      2. Aesthetics

3. Ethical Considerations

4. Evolutionary Potential

5. Benefits to Agriculture, Medicine, and Industry

IV. Managing Biological Resources

   A. How Have Biological Resources Been Managed Historically?

   B. How Are Biological Resources Currently Managed?

   1. Off-Site Management of Plants

   2. Off-Site Management of Animals

   3. On-Site Management of Plants and Animals

   4. Pros and Cons of Off-Site and On-Site Preservation

   C. What Is the Endangered Species Act?

   1. Species listed under the ESA may not be harassed, harmed, hunted, trapped or killed, nor may they be bought, sold, imported or exported.

   D. What Is CITES?

   E. What Is the Convention on Biological Diversity?

Learning Objectives

After learning the material in Chapter 17, you should be able to:

1. Define biological diversity. Differentiate between genetic diversity, species diversity, and ecosystem diversity.

2. Define extinction and explain the relationship between extinction and biological diversity.

3. Identify the major threats to biological diversity.

4. Explain why it is important to preserve biological diversity.

5. Briefly relate how plant and animal species have been managed historically.

6. Describe how species are currently managed both within and outside of their natural habitats, and explain the advantages and disadvantages of each.

7. Discuss the major legislation, national and international, to protect and preserve biodiversity.

Did You Know...?

Tropical rain forests, while covering only seven percent of the Earth's surface, contain nearly half of all known species. However, much of this biodiversity may be lost as tropical rain forests have already surrendered nearly one-half of their original area to deforestation.

 ## Key Concepts

Read this summary of Chapter 17 and identify the important concepts discussed in the chapter.

All species are biological resources. Biological diversity refers to the variety of life forms which inhabit the Earth. Biological diversity is measured in terms of species diversity, the total number of species; ecosystem diversity, the variety of communities of organisms and their habitats; and genetic diversity, the variation among the members of a single population of a species. Each individual has a unique genotype, its composition of genes. The sum of all of the genes present in a population is a gene pool.

The number of species on Earth today is unknown. Scientific estimates range from three million to 100 million. Only about 1.7 million species have actually been named, and just three percent of these have been studied. Also unknown is the rate of extinction, the elimination from the biosphere of all individuals of any taxonomic rank (family, genus, or species). While extinction is a natural process, human activities have greatly accelerated its rate in recent decades.

The planet's diverse ecosystems harbor millions of species. Endemic species are species unique to a particular ecosystem; they occur nowhere else on Earth. The planet's most diverse ecosystems are the tropical rain forests and the coral reefs of the oceans. The former is threatened by deforestation, conversion to agricultural land, roadbuilding, mining, and other development. Reefs are threatened by dynamiting, coral and shell collecting, siltation from streams and rivers flowing into the seas, and bleaching, which some researchers believe may be a result of global warming. Wetlands are also very diverse, and like rain forests and coral reefs, they are under great pressure, chiefly from development.

Social factors, particularly culturally inherited attitudes and beliefs about wild species, affect how we value, use, and manage biological resources. For example, endangered mammals or birds that are highly valued by humans are called "charismatic megafauna." Undesirable animals and plants are called "vermin" and "weeds," respectively. Species hunted for sport are game animals, while species not hunted for sport are nongame animals. Most monies are directed at managing and protecting game animals and charismatic megafauna. Meanwhile, efforts to manage nongame animals and less "cuddly" species (like snakes or tortoises, for example) receive little funding.

Our attitudes and beliefs also give rise to many activities which threaten biological resources and biodiversity. The single greatest threat to wildlife is loss or degradation of habitat. Other activities that endanger biological resources are overharvesting and illegal trade, selective breeding, introduced species, and pollution. Why preserve diversity? The arguments can be grouped into five broad categories: ecosystem services; aesthetics; ethical considerations; evolutionary potential; and benefits to agriculture, medicine, and industry.

Humans have been managing plant and animal species for thousands of years. From China to Egypt to South America, the collection of wild animals and plants by rulers and the elite were fairly common. Zoos originated in Europe, an off-shoot of game reserves. Their chief purpose was entertainment. Current efforts to manage biological resources concentrate on preserving plant and animal species or their genetic material, known as germ plasm. Both species and germ plasm may be preserved outside of their natural habitat in private collections, wildlife refuges, zoological parks, and seed or gene banks. Those who attempt to breed wild animals in captivity strive to maximize the contribution of unrelated animals in order to reduce the effects of inbreeding, and to attain a good age distribution in order to ensure a consistent number of individuals of breeding age.

There are numerous advantages to off-site preservation. For many species, preservation in a zoo or botanical garden offers its best or only hope for long-term survival. In some cases, species may be preserved off-site until a time when they can be reintroduced to the wild. Off-site institutions also play a vital educational role; both adults and children learn about wildlife mainly through zoos. The most obvious disadvantage of off-site preservation is that it is an alternative only for those species known to us. It is also very expensive, and only a relatively small number of species can be preserved in zoos, botanical gardens, and gene banks because of space limitations. Further, in their natural habitat, species generally adapt to environmental changes. Sustained preservation of a species off-site reduces the likelihood of its successful reintroduction to the wild. In addition, we lack the basic biological knowledge needed to effectively manage and breed many exotic species. Finally, preserving species off-site ignores the other components of their ecosystems.

On-site preservation is accomplished through the establishment of national parks, protected wilderness areas, and biosphere reserves. It offers numerous advantages. Chief among these is that on-site preservation protects the entire ecosystem, both known and unknown species and the relationships between them. It holds the best hope for preserving maximum biological diversity. On-site preservation also allows a species to evolve with its environment. It is generally less expensive than off-site preservation, although protecting the designated preserve against poachers and development can greatly increase the cost.

Historically, most management efforts in the United States were aimed at game species, such as large mammals hunted for their horns or skins (furs) or large birds killed for their plumage. Most legislation dealt with the trade or transport of wild animals or animal products and thus did not protect living animals and their habitats. Moreover, such legislation ignored threatened plant species. The Endangered Species Act (ESA) of 1973 marked a change in philosophy; it acknowledged the need to protect a diversity of species (plant and animal) as well as habitats.

Weaknesses of the ESA include the following: listing a species as threatened or endangered is expensive and time-consuming; there are not enough monies provided to adequately study and list all rare species; priority is given to mammals and more highly-visible creatures; and the listing process can be circumvented by the "God Committee," which can reverse the U.S. Fish and Wildlife Service's decision to protect endangered species.

At the international level, many countries have adopted various international treaties and conventions to protect migratory species. For the most part, these treaties have been bilateral agreements and have involved only a few countries in a specific region. One treaty, however, is global in nature. CITES, the Convention on International Trade in Endangered Species of Wild Flora and Fauna, protects threatened plants and animals worldwide and governs international trade in wildlife and wildlife products. Some weaknesses of CITES are that some countries have not joined CITES and thus do not adhere to its decisions; signatory countries can ignore a CITES injunction simply by taking out a "reservation" on a species; and enforcement varies widely from country to country.

Key Terms

biodiversity	biological diversity
biological resources	charismatic megafauna
charismatic minifauna	ecosystem diversity
endemic species	extinction
game animal	gene
gene pool	genetic diversity
genetic erosion	genotype
germ plasm	minimum viable population size
nongame animal	species diversity
vermin	weed

Environmental Success Story

Cattle grazing has long been assumed to be at odds with environmental concerns. Cattle hooves compact soil, which can lead to runoff; overgrazing can trigger erosion; and large-scale ranching can displace native flora and fauna, thus reducing biological diversity. But when The Nature Conservancy's Wyoming Chapter acquired Red Canyon Ranch in 1993, workers turn the seemingly destructive aspects of cattle ranching into advantages for both humans and nature. Through a program of active grazing management, cows' hooves now plant seeds and stomp out weeds; the animals' dung and urine are used to fertilize native plants; and their movements are restricted to protect local streams. The results are impressive: Native grasses and plants have returned to the ranch, nearby streams have been restored, and local habitats are flourishing with populations of songbirds, moose, deer, beaver, elk, antelope, and bighorn sheep.

Source: The Nature Conservancy. "Eighth Annual National Awards for Environmental Sustainability Winners: Urban Red Canyon Ranch." *Renew America* (22 April 1998). Available: http://solstice.crest.org/sustainable/renew_america/winner98.html. 24 March 1999.

True/False

1. Endemic species are species that are unique to a certain area and occur nowhere else in the world.   T   F

2. Wetlands, because they are transitional zones between land and water, tend to be relatively low in species diversity.   T   F

3. About 10 percent of the Earth's land surface is protected habitat (such as parks and reserves).   T   F

4. CITES prohibits the international trade of hundreds of endangered species.   T   F

5. There are approximately 1.4 million species on Earth.   T   F

Fill in the Blank

1. The variation among members of a single population is called _____.

2. _____ is a term for loss of genetic variability in a population.

3. The use of _____ instead of traditional or wild strains contributes to the loss of genetic diversity in agriculture.

4. The _____ are typically large, attractive mammals which are popular with the general public; their management and protection are well-funded.

5. _____ have been called the rain forest of the oceans because of the abundance and diversity of life found there.

Multiple Choice

Choose the best answer.

1. The hereditary material of an organism is known as its
    A. gene pool.
    B. genetic variation.
    C. genotype.
    D. germ plasm.

2. An endemic species is one that
    A. occurs in only one area.
    B. cannot survive in simulated natural habitats.
    C. is introduced to an area in which it not native.
    D. None of the above is true.

3. The most serious threat facing wildlife is
    A. poaching.
    B. overharvesting.
    C. habitat loss.
    D. selective breeding.

4. Zebra mussels in the Great Lakes are an example of a(n)
    A. charismatic megafauna.
    B. endangered species.
    C. endemic species.
    D. exotic species.

5. The dominant type of coastal wetland in tropical and subtropical regions is the
    A. bog.
    B. mangrove swamp.
    C. estuary.
    D. oxbow.

Short Answer

1. How many species inhabit the Earth?
2. Which two types of ecosystems have the greatest species diversity?

3. Explain the terms "charismatic megafauna" and "vermin." To what do they refer and what does each term imply?

4. List and briefly describe three ways that zoos group animals.

5. What is ISIS?

Thought Questions

Develop a complete answer for each of the following.

1. Discuss how human culture and beliefs affect which species are valued and protected.

2. Describe how human activities threaten biological diversity.

3. Explain the arguments for preserving biological resources and maintaining biological diversity.

4. How do the zoos of today differ from the zoos of the past? Discuss how the reasons for collecting plant and animal species have changed over time.

5. Compare off-site and on-site preservation. Explain their differences in terms of species, ecosystem, and genetic diversity.

6. Describe the challenges faced by captive breeding programs.

7. Discuss the strengths and weaknesses of U.S. legislation designed to preserve biological diversity.

■ Related Concepts

Describe the relationship. (There may be more than one.)

BETWEEN...	AND...
extinction	ecosystem diversity
genetic diversity	species diversity
international wildlife trade	CITES
species	ecosystems

Did You Know...?

More than 1,700 species of birds inhabit the rain forests in Colombia alone, while only 700 species can be found on the entire North American continent.

Suggested Activities

1. Find out what your local zoo is doing to preserve biological diversity. Volunteer to help with an activity or event that intrigues you.

2. Put up a bird house or bird bath. Observe the species that visit.

3. Start a butterfly garden; be sure to include a variety of flowers and other plants that attract a wide range of butterflies — like milkweed for Monarch butterflies.

UNIT **V**

An Environmental Pandora's Box:
Managing the Materials and Products of Human Societies

Chapter 18
Mineral Resources

 Chapter Outline

I. **Describing Mineral Resources: Physical and Biological Characteristics**

   A. How Did the Earth Form and How Does It Change?

      1. Magma

      2. Igneous Rock

      3. Sedimentary Rock

      4. Metamorphic Rock

   B. What Are Minerals?

      1. A mineral is a nonliving, naturally occurring substance with a limited range in chemical composition and with an orderly atomic arrangement.

   C. How Are Minerals Classified?

      1. Fuels

      2. Nonfuels

         a. Metallic

         b. Nonmetallic

            i. Ferrous

            ii. Nonferrous

      3. Abundant Metals

      4. Scarce Metals

   D. How Are Mineral Deposits Formed?

   E. Where Are Minerals Found?

      1. Minerals are found everywhere but are typically bound up in rocks within the Earth's crust.

II. **Describing Mineral Resources: Social Characteristics**

   A. How Are Minerals Used?

      1. The more-developed countries (MDCs) consume the greatest share of mineral resources.

B. How Are Mineral Deposits Classified?

1. Resource Base

2. Proven Reserve

3. Subeconomic Resources

4. Ultimately Recoverable Resource

C. What Are the Steps in the Mining Process?

1. Location

2. Extraction

a. Surface mining

i. Open-pit surface mining

ii. Area strip mining

iii. Contour strip mining

b. Subsurface mining

3. Processing

4. Summary: Environmental Impact of Mining

III. Managing Mineral Resources

A. What Is the Historical Significance of Minerals?

B. How Do Economic Factors Affect Mineral Production and Consumption?

C. What Is the International Minerals Industry?

D. How Is Seabed Mining Overseen?

E. How Does the United States Manage Mineral Supplies?

1. Domestic Mineral Supplies

2. Imported Mineral Supplies

3. Conservation Of Mineral Resources

Learning Objectives

After learning the material in Chapter 18, you should be able to:

1. Explain how mineral deposits formed in the Earth's crust.

2. Distinguish between proven and recoverable mineral resources.

3. Describe the steps and environmental consequences of the mining process.

4. Compare mineral availability and use in MDCs and less-developed countries (LDCs).

> **Did You Know...?**
>
> Don't assume a product is toxic-free because there are no toxins listed on the label. The government doesn't require manufacturers to list every ingredient if it doesn't violate Federal Safety Standards. Baby powder, for example, often contains asbestos. Traces of pesticides have been found in shampoos.

 ## Key Concepts

Read this summary of Chapter 18 and identify the important concepts discussed in the chapter.

Although inorganic matter may seem to be unchanging, it has actually undergone much transformation since the Earth began — and it continues to change. Igneous rock is formed through volcanic activity. Sedimentary rock is formed by the deposit of small bits and pieces of matter, or sediments, that are carried by wind or rain and then compacted and cemented to form rock. Metamorphic rock forms when rocks lying deep below the Earth's surface are heated to such a degree that their original crystal structure is lost. Different crystals form as the minerals that compose the rock cool.

The Earth consists of several different layers. At the center is a dense, metallic core, which is surrounded by a mantle. The outer layer is called the crust and contains the mineral deposits exploited by humans.

Minerals are nonliving, naturally occurring substances with a limited range in chemical composition and with an orderly atomic arrangement. A mineral may occur in one of three forms: as a single element, a compound of elements, or an aggregate of elements and compounds. Each mineral's unique chemical composition determines its physical properties, such as strength, insulating or sealing capacity, electrical conductivity, or beauty. Minerals are broadly classified as fuels or nonfuels. Nonfuels are further classified as metallic or nonmetallic. Metals are sometimes classified by their abundance or scarcity in the Earth's crust. Abundant metals make up more than 0.1 percent of the Earth's crust, by weight. Scarce metals make up less than 0.1 percent of the Earth's crust.

Minerals can be found everywhere, but they are not evenly distributed throughout the Earth's crust. Many scientists believe the distribution of mineral deposits is related to the past and present movement of the Earth's tectonic plates. Additionally, deposits of one mineral are often mixed with deposits of others. Mineral deposits are formed by chemical separation and are affected by gravity. They are dispersed by weathering and erosion.

To be useful, a mineral must be profitable to extract. Where a concentration is high enough to make mining economically feasible, the mineral deposit is known as an ore. Some nations are mineral-rich and others are mineral-poor.

Minerals are used in many different ways: in steel production, manufacturing, industrial processes, fertilizers, and construction. The greatest share of mineral resources is consumed by the MDCs. MDCs, with about one-quarter of the world's population, use about three-quarters of the global production of nonfuel minerals. Global demand for most major minerals is expected to rise significantly as the human population continues to grow rapidly and societies continue to rely on mineral resources to support rising standards of living. MDCs rely on about 80 minerals; three-quarters of these either exist in abundant supply to meet anticipated needs or can be replaced by existing substitutes. Critical minerals are those considered essential to a nation's economic activity; strategic minerals are those considered essential to a nation's defense. At present, there are no suitable alternatives for critical or strategic minerals.

An estimate of the total sum of a mineral found in the Earth is called the resource base; this estimate is highly theoretical. A mineral deposit that can be extracted profitably with current technology is called a proven reserve or economic resource. Subeconomic resources are reserves that have been discovered but cannot yet be extracted at a profit at current prices or with current technology. An estimate of the total amount of a given mineral that is likely to be available for future use is called the ultimately recoverable resource. This estimate is based on known reserves, plus assumptions about discovery rates, future costs, market factors, and future advances in extraction and processing technologies. As reserves become depleted, the cost is likely to rise. High prices and high demand encourage more exploration and lead to the discovery of more ores or the development of suitable substitutes.

Mining includes three major steps: location, extraction, and processing. All three can adversely affect human and environmental health. Modern exploration relies on knowledge of geology and the use of sophisticated equipment. Most high-grade ores have already been identified and exploited. The ocean floor and Antarctica hold real potential for profitable mining, but because they are global commons, disputes over mining claims are problematic. Extraction is the process of separating a mineral ore from the surrounding rock in which it is embedded. Ore may be extracted by surface mining or subsurface mining techniques. Extraction degrades the environment by promoting erosion and siltation, rearranging the layers of soil, leaving tailings that contain hazardous substances, and polluting water. Extraction is also hazardous to human health. Mineral processing consists of separating the mineral from the ore in which it is held and concentrating and refining the separated mineral. Processing ores causes soil and water pollution and also affects human health.

Market supply and demand determine the price of minerals. Supply and demand, in turn, are influenced by the needs and uses of society, technological constraints, economic forces, and resource scarcity. Scarcity is related to political, economic, and technical factors, not actual reserves. Mineral prices tend to fluctuate wildly in the short term, but over time the economic value attached to commodity exports such as minerals has fallen relative to the value of manufactured goods. Countries that rely on mineral exports, particularly LDCs, must sell increasing amounts of minerals to pay for imports such as tractors and fertilizers.

In the United States, the development of domestic mineral deposits on public land is governed by the Mining Law of 1872, which grants title to certain public lands as long as certain criteria are met. Some mineral deposits are currently governed under the 1920 Mineral Leasing Act, which allows private individuals and companies to lease the rights to develop these resources. In the late 1990s, the debate between mineral development and preservation on the public lands intensified with a bill designed to address the weak areas of the Mining Law of 1872. The focus of the debate on wilderness designation and permanent protection for national parks is shifting to Alaska, which contains sizable reserves of many important nonfuel minerals. The United States imports most strategic and critical minerals. To guard against shortages, it stockpiles important minerals.

Minerals can be conserved by finding substitutes for those that are in short supply, reusing products and materials, and recycling materials. Advanced materials are substances that can replace traditional materials and have more of a desired property. They generally use less metals and other minerals than traditional materials or include minerals that have not been widely used. The "throwaway mentality" of American society has caused us to recycle significantly less than we are able to. This attitude must change if we are to conserve material resources and reduce solid waste.

 ## Key Terms

abundant metal	advanced material
area strip mining	contour strip mining
critical mineral	economic resource
ferrous metal	igneous rock
magma	metamorphic rock
mineral	mineral extraction
mineral policy	mineral processing
nonferrous metal	open pit surface mining
ore	overburden
proven reserve	resource base
scarce metal	sedimentary rock
strategic mineral	subeconomic resource
subsurface mining	surface mining
tailings	ultimately recoverable resource

Environmental Success Story

Many regions in the United States are searching for ways to expand their existing economic base with environmentally friendly industries. North Carolina's HandMade in America (HIA) has found one unique way: It promotes the area's handcraft industry, an alternative to the mining, logging, and manufacturing that predominates the region. The creation of handcrafts, such as furniture, glass, ceramics, jewelry and leather products, offers reduced environmental impact while generating substantial revenue. A recent study shows that the western North Carolina craft industry directly contributes $122 million annually to the region's economy, and millions more indirectly through related tourism. HIA promotes the handcraft industry through a program that offers low-interest loans to craftspeople and brokering agreements with farmers' markets to feature agriculturally based craft products. With the National Trust for Historic Preservation, HIA also created a "national craft heritage corridor," featuring historic craft sites, scenic byways, and retail shops and galleries carrying area crafts. HIA is showing the nation one way to protect the environment while creating long-term economic growth.

Source: HandMade in America. "Eighth Annual National Awards for Environmental Sustainability Winners: HandMade in America." *Renew America* (22 April 1998). Available: http://solstice.crest.org/sustainable/renew_america/winner98.html. 24 March 1999.

True/False

1. Ferrous metals contain either iron or elements alloyed with iron to make steel.   T   F
2. Manganese and magnesium are abundant metals.   T   F
3. Most ores in the U.S are extracted through subsurface mining.   T   F
4. The U.S. contains adequate deposits of most strategic and critical minerals.   T   F
5. Mineral scarcity is primarily determined by the amount of known reserves.   T   F

Fill in the Blank

1. Minerals are formed through geologic processes involving the cooling of _____ from the Earth's core.
2. Where the concentration of a mineral is high enough to make mining economically feasible, the deposit is known as a(n) _____.
3. _____ is the vegetation, soil, and rock removed during surface mining.
4. A mineral considered essential to a nation's defense is called a(n) _____.

5. To provide a reserve of strategic and critical minerals in the event of a sudden cutoff or embargo, the U.S. _____ these minerals.

Multiple Choice

Choose the best answer.

1. Which of the following is true of U.S. mineral consumption?

    A. The U.S. consumes more iron than any other country.

    B. The U.S. consumes about 30 percent of total global mineral production (fuel and nonfuel).

    C. The U.S. consumes fewer critical mineral resources than most MDCs.

    D. None of the above is true.

2. Mineral resources which have been discovered but cannot yet be extracted at a profit are known as

    A. proven resources.

    B. subeconomic resources.

    C. recoverable resources.

    D. economic resources.

3. The most environmentally benign step in the mining process is

    A. exploration.

    B. extraction.

    C. processing.

    D. All of the above are equally damaging.

4. For most minerals, stockpile targets are set at a _____ year supply.

    A. 1

    B. 3-4

    C. 10

    D. 20-25

5. Which of the following is an application of advanced materials?

    A. communications industry

    B. automotive

    C. packaging

    D. All of the above are true.

Short Answer

1. List and briefly describe the three types of rocks.
2. List and briefly describe the different classifications of minerals.
3. Explain the difference between critical and strategic minerals.
4. What is the Law of the Sea?
5. What advantages do advanced materials offer?

Thought Questions

Develop a complete answer for each of the following.

1. How are mineral deposits formed?
2. Discuss estimates of mineral deposits. Why is it unlikely that we will exhaust mineral resources?
3. Describe the steps in the mining process and discuss the environmental effects of each step.
4. Discuss the economic and environmental implications of mining deposits in Antarctica and the oceans.
5. Discuss the factors that affect mineral production and consumption.
6. Describe the conflict between mining laws and environmental conservation in the U.S.
7. How can mineral resource supplies be conserved?

▪ Related Concepts

Describe the relationship. (There may be more than one.)

BETWEEN...	AND...
mineral distribution	tectonic activity
geology	mineral exploration
tailings	overburden
Mining Law of 1872	Mineral Leasing Act of 1920
advanced materials	conservation of mineral resources

Did You Know...?

American consumers and industries throw away enough aluminum to rebuild their entire commercial air fleet every three months.

 Suggested Activities

1. Learn about ways in which you can recycle materials, and begin recycling. Encourage your friends and family to do the same. Find out what happens to the materials you recycle.

2. Write your Congressional representatives to voice your opinion on current plans to revise mining laws.

3. Visit a mining site. Try to arrange a tour so that you can learn more about how minerals are extracted and processed and about how mining sites are reclaimed.

Chapter 19
Nuclear Resources

 Chapter Outline

I. **Describing Nuclear Resources: Physical Characteristics**

   A. What Are Nuclear Resources?

      1. Nuclear Energy

      2. Radiation

      3. Ions

      4. Ionizing Radiation

      5. Nonionizing Radiation

   B. How Is Nuclear Energy Released?

      1. Spontaneous Radioactivity

         a. Radioactive decay

         b. Isotopes

         c. Radioisotopes

         d. Alpha particles

         e. Beta particles

         f. Gamma radiation

         g. X-rays

         h. Half-life

      2. Fission

         a. Fission products

         b. Critical mass

         c. Chain reaction

      3. Fusion

         a. Fusion reaction

II. **Describing Nuclear Resources: Biological Characteristics**

   A. How Does Radiation Affect Human Health?

1. Amount of Ionizing Radiation
2. Age and Gender

B. How Does Radiation Enter and Affect the Environment?

1. Direct Exposure

    a. Radioactive fallout

2. Indirect Exposure

    a. Nuclear winter

III. Describing Nuclear Resources: Social Characteristics

A. How Are Nuclear Resources Used?

1. Medical Applications
2. Generation of Electricity
3. Food Preservation
4. Military Applications

IV. Managing Nuclear Resources

A. How Has the Use of Nuclear Resources Changed?

1. Military Uses of Nuclear Energy

    a. The Cold War and the Arms Race

    b. Attempts at arms reduction

2. Peaceful Uses of Nuclear Energy

B. What Environmental Problems Are Associated with the Use of Nuclear Resources?

1. Radioactive Leaks
2. Accidents at Nuclear Power Plants

    a. Meltdowns

3. Disposal of Nuclear Wastes

    a. Uranium mill tailings

    b. Low-level radioactive waste

    c. Transuranic radioactive waste

    d. High-level radioactive waste

4. Decommissioning Old Plants
5. Secrecy Surrounding Nuclear Activities

Learning Objectives

After learning the material in Chapter 19, you should be able to:

1. Describe the different applications of nuclear technology.
2. Explain two types of exposure to radiation and discuss the possible health effects.
3. Identify four types of waste produced by a nuclear power plant.
4. Identify the issues involved in long-term storage of nuclear waste.

Did You Know...?

Between 1940 and 1996, the U.S. government spent an estimated $5.5 trillion on the nuclear arms race. If divided equally among all of today's Americans, the cost would be $22,000 per person.

Key Concepts

Read this summary of Chapter 19 and identify the important concepts discussed in the chapter.

Nuclear resources are derived from atoms, their energy, and the particles the emit. Nuclear energy is the energy released, or radiated, from an atom. Energy released from an atom is called radiation. Radiation takes two basic forms: ionizing and nonionizing.

Ionizing radiation travels in waves (x-rays or gamma rays) or as particles (alpha particles or beta particles). The energy level of ionizing radiation is high enough to remove electrons from atoms, creating charged particles called ions. Nonionizing radiation (heat, light, radiowaves) can also affect atoms, but its energy level is not high enough to create ions.

Radioisotopes are those isotopes (atoms of the same element which have different numbers of neutrons) that release particles or high-level energy. Nuclear energy is released through three types of reactions: spontaneous radioactivity, fission, and fusion. Spontaneous radioactivity occurs when unstable atoms release mass in the form of particles (particulate radiation), energy in the form of waves (electromagnetic radiation), or both. This release of energy and/or particles is known as radioactive decay. The length of time it takes for any radioactive substance to lose one-half of its radioactivity is known as its half-life.

A fission reaction occurs when an atom is split into two or more new atoms. The atoms formed when uranium in a nuclear reactor is split are called fission products. The smallest amount of fuel necessary to sustain a chain reaction is known as the critical mass. A chain reaction is a self-perpetuating series of events that occur when a neutron splits a heavy atom, releasing additional neutrons to cause other atoms to split.

A fusion reaction is the opposite of a fission reaction; it occurs when nuclei are forced to combine, or fuse. For many people, fusion holds the promise of "clean" nuclear energy, but it is not yet feasible for commercial electric generation.

The most common types of ionizing radiation are alpha, beta, and gamma radiation. Alpha particles consist of two protons and two neutrons and carry a positive charge. Beta particles are negatively charged particles emitted from nuclei. Gamma radiation is a powerful electromagnetic wave. X-rays are a form of cosmic radiation; x-rays can also be produced by firing electrons at tungsten metal.

Ionizing radiation adversely affects human health. As it penetrates living tissue, ionizing radiation can destroy cells or alter their genetic structure. If the dose of radiation is high enough and the resulting damage too severe, the cell will die. Cells receiving nonlethal doses may exist with altered DNA for years, spawning cancerous cells and eventually tumors. Over time, the cellular change caused by ionizing radiation can lead to a wide variety of health problems including cancers, cataracts, mental retardation, and a weakened immune system. The effect of radiation on human beings is dependent upon the amount of exposure they receive, as well as age and gender.

Humans can be exposed to radiation in the environment in one of two ways. Direct exposure is exposure to the original radioactive source (a power plant suffering a nuclear leak, for example) or to dirt and debris contaminated with radiation, material known as radioactive fallout. Indirect exposure is exposure to radioactive substances through food chains.

Nuclear resources are used in a variety of ways, including medical applications, electric generation, food preservation, and military applications.

During World War II, the military took the initial lead in developing nuclear resources. The bombings of Hiroshima and Nagasaki demonstrated the awesome destructive potential of the atom. In 1947, the government established the Atomic Energy Commission (AEC) to control the use and disclosure of information on atomic power. When the Soviet Union detonated its first atomic bomb in 1949, the nuclear arms race between that country and the United States began. The Cold War, as it is known, ended with the break-up of the Soviet Union; even so, the United States and Russia still hold significant nuclear arsenals. Other countries known to possess nuclear arsenals are France, Britain, and China. Israel, India, and Pakistan are also believed to have nuclear capabilities.

The use of nuclear resources has been accompanied by many environmental and social problems. The most serious of these are radioactive leaks, accidents at nuclear power plants, disposal of radioactive wastes, decommissioning of old plants, and the secrecy surrounding nuclear activities.

✓ Key Terms

alpha particle	beta particle
chain reaction	containment vessel
critical mass	direct radiation exposure
fission product	fission reaction
fusion reaction	gamma radiation
half-life	high-level radioactive waste
indirect radiation exposure	ionizing radiation
ions	isotope
low-level radioactive waste	meltdown
nonionizing radiation	nuclear energy
nuclear resources	nuclear winter
radiation	radioactive decay
radioactive fallout	radioisotope
spontaneous radioactivity	transuranic radioactive waste
uranium mill tailings	x-ray

Environmental Success Story

The City of Los Angeles is meeting the challenge of managing seven pounds of trash per person each working day. With 3.5 million people spread over 450 square miles, the United States's second largest city has successfully managed solid waste through innovation and community participation. By combining automated trash collection and co-mingled recycling, the city is making recycling and waste disposal convenient, clean, and easy for residents. Each month, between 4,000 and 5,000 tons of recyclables are diverted and redirected to facilities for recovery. In addition, 25,000 tons of yard trimmings are composted and sold to farmers in the San Joaquin Valley, or bagged and sold in Los Angeles gardening and home stores. Boasting the largest curbside recycling program in North America, the city also has created waste management strategies that strengthen the recyclables market, train and employ inner-city youth, improve conditions for sanitation workers, and better the quality of life for all residents.

Source: Los Angeles Resource Program. "Eighth Annual National Awards for Environmental Sustainability Winners: City of Los Angeles Board of Public Works." *Renew America* (22 January 1997). Available: http://crest.org/sustainable/renew_america//winners.html. 18 February 1999.

True/False

1. Most of the radiation released to the environment comes from natural sources.   T   F
2. Doses of over five rems per year are considered high-level exposure.   T   F
3. According to the Nuclear Regulatory Commission, the maximum safe exposure for the general public is 0.17 millirem per year.   T   F
4. The 1963 Nuclear Test Ban Treaty halted above-ground testing.   T   F
5. Nuclear power produces only about five percent of the total electricity produced in the U.S.   T   F

Fill In The Blank

1. The period it takes for half of the atoms of a radioactive material to decay into the next element is its _____.
2. Dirt and debris contaminated with radiation is called _____.
3. The radioactive element cobalt 60 is used in the process of _____ to kill bacteria, insects, and fungi on produce.
4. The most familiar use of radiation in therapeutic medicine is in the treatment of _____.
5. When a nuclear power plant becomes to old to function, it is _____ by dismantling and decontaminating the reactor.

Multiple Choice

1. Ionizing radiation
    A. travels in particles and waves.
    B. is produced by fission reactions.
    C. is capable of damaging living tissue.
    D. All of the above are true.
2. Negatively charged particles that contain more energy that electrons are
    A. alpha particles.
    B. beta particles.
    C. gamma rays.
    D. x-rays.

3. The amount of radiation absorbed per gram of tissue is expressed in units called

   A. rads.

   B. rems.

   C. sieverts.

   D. None of the above is true.

4. A problem associated with older nuclear power plants is

   A. corroded fuel rods.

   B. brittle steel pressure vessels.

   C. pressurized reactor core.

   D. All of the above are true.

5. Waste that contains human-made radioactive elements with an atomic number higher than uranium is called

   A. low-level radioactive waste.

   B. high-level radioactive waste.

   C. transuranic radioactive waste.

   D. None of the above is true.

Short Answer

1. What is a nuclear winter?
2. Where was the world's first commercial nuclear reactor built?
3. What were the stipulations of the 1968 Nuclear Non-Proliferation Treaty?
4. What is a meltdown?
5. What does NIMBY stand for?

Thought Questions

1. Describe what happens in a fusion reaction. What are the barriers to producing man-made fusion reactions? Is fusion the answer to the problems caused by fission reactions?

2. Discuss the known effects of radiation on human health. Explain what is meant by low-level and high-level doses. What factors seem to affect susceptibility?

3. How do we use nuclear resources? What are the major risks? What are the strongest arguments for and against using nuclear resources?

4. What are the environmental effects of using nuclear energy? How does it compare with other energy sources?

5. Describe the evolution of our use of nuclear resources, from the Manhattan Project up to the present day. What has been the U.S. government's role in the development of nuclear resources?

6. Explain the methods for disposing different kinds of nuclear wastes and describe the problems associated with each.

■ Related Concepts

Describe the relationship. (There may be more than one.)

BETWEEN...	AND...
fission	fusion
indirect exposure	radioactive rain
low-level radioactive waste	high-level radioactive waste
Atomic Energy Commission	Department of Energy

Did You Know...?

Worldwide, enough hazardous waste is generated in one year to fill the New Orleans Superdome 1,500 times over.

 Suggested Activities

1. Find out if your home receives energy from a nuclear power plant or plants. If so, investigate it. How long has it been in operation? What is its safety record?

2. Do some more research on Yucca Mountain, Nevada, the site selected by the Department of Energy for storage of high-level radioactive waste. What is planned for the site? What are the barriers to its use?

3. Test your home for radon gas.

Chapter 20
Toxic and Hazardous Substances

 **Chapter Outline**

I. **Describing Toxic and Hazardous Substances: Physical Characteristics**

   A. What Are Toxic and Hazardous Substances?

      1. Chemicals that can adversely affect human health and the environment

   B. What Is the Difference Between Toxic and Hazardous Substances?

      1. The term "toxic" implies a potential to cause injury to a living organism.

      2. The term "hazardous" implies there is some chance that an organism will be exposed to a toxic substance and that exposure will result in harm.

   C. How Do Toxic and Hazardous Substances Enter the Environment?

      1. As By-Products

      2. As End Products

II. **Describing Toxic and Hazardous Substances: Biological Characteristics**

   A. How Do Toxic and Hazardous Substances Affect Human and Environmental Health?

      1. Acute Toxicity

      2. Chronic Toxicity

      3. Carcinogenic Substances

      4. Infectious Substances

      5. Teratogenic Substances

      6. Mutagenic Substances

      7. Persistent Organic Pollutants

         a. Endocrine disrupters

III. **Describing Toxic and Hazardous Substances: Social Characteristics**

   A. Who Produces and Uses Toxic and Hazardous Substances?

      1. Worldwide, production and use of toxic substances are concentrated in the industrialized nations, although less-developed countries use significant quantities of pesticides.

2. Most of the world's hazardous wastes are generated and disposed of by the chemical, primary metals, and petroleum-refining industries.

B. What Are Household Hazardous Wastes?

C. How Do Toxic and Hazardous Substances Affect Communities?

1. Environmental Racism

2. Transport of Toxic and Hazardous Materials

IV. Managing Toxic and Hazardous Substances

A. What Is Incineration?

1. Burning of Wastes

B. What Is Landfilling?

1. Sanitary Landfills

2. Secure Hazardous Waste Landfills

C. What Is Deep-Well Injection?

1. Pumping liquid waste deep underground into permeable rock formations that geologists believe will contain the waste permanently

D. How Are Toxic and Hazardous Substances Treated to Reduce the Risk of Environmental Contamination?

1. Phase Separation Processes

2. Component Separation Processes

3. Solidification

4. Bioremediation

E. What Is Pollution Prevention?

1. Source Reduction

2. Waste Minimization

F. What Are Waste Exchanges?

1. The process of locating and bringing together companies that have wastes and companies that want to recover and reuse these resources

G. What Legislation Affects the Management of Toxic and Hazardous Substances?

1. Superfund: The Comprehensive Environmental Response, Compensation, and Liability Act (CERCLA)

2. Resource Conservation and Recovery Act

a. Regulating hazardous wastes

b. Cradle-to-grave management of hazardous wastes

3. Toxic Substances Control Act

Learning Objectives

After learning the material in Chapter 20, you should be able to:

1. Give some examples of toxic and hazardous substances and how they enter the environment.
2. Describe three different methods of disposing of hazardous waste.
3. Describe the concept of pollution prevention and explain why it is important.
4. Identify three major pieces of U.S. legislation that regulate toxic and hazardous substances.
5. Identify the locations of two major chemical disasters.

Did You Know...?

More than 150 toxic waste dumps line the banks of the Niagara River in one three-mile stretch.

 ## Key Concepts

Read this summary of Chapter 20 and identify the important concepts discussed in the chapter.

Toxic and hazardous substances are chemicals that can adversely affect human health and the environment. This broad definition can include elements (such as lead), compounds (such as polychlorinated biphenyls, or PCBs), and the products of infectious agents like bacteria and protozoans. Many toxic and hazardous substances occur naturally in the Earth's crust and biota; others are manufactured in industrial processes.

Although the terms "toxic" and "hazardous" are often used interchangeably, they are not synonymous. The term "toxic" implies the potential to cause injury to living organisms. Almost any substance can be toxic under the right conditions, if the concentration is high enough or if an organism is exposed to the substance for long enough. The term "hazardous" implies that there is some chance that the organism will be exposed to a substance and that exposure will result in harm. Substances are considered hazardous when a possibility exists that plants and animals will be exposed to them.

The 1976 passage of the Resource Conservation and Recovery Act (RCRA) provided the EPA with a statutory framework for defining hazardous waste. Accordingly, a hazardous waste is defined as any solid, liquid, or gaseous waste which, due to its quantity, concentration, or physical, chemical or infectious characteristics, may cause or significantly contribute to an increase in mortality or serious illness or pose a substantial present or potential hazard to human health or the environment when improperly stored, transported, disposed of, or recycled.

Toxic and hazardous substances are released into the environment as by-products, as end products, or through the use and disposal of manufactured products. They are of serious concern because of their potential and suspected adverse effects on ecosystems and living organisms, including humans.

Toxic and hazardous substances affect human and environmental health in a variety of ways. Acute toxicity is the occurrence of serious symptoms immediately after a single exposure to a substance. Chronic toxicity is the delayed appearance of symptoms until a substance accumulates to a threshold level in the body after repeated exposures to the substance. There are also a number of general ways in which hazardous and toxic substances affect long-term human health. Carcinogenic substances can cause cancer in humans and animals. Infectious substances contain disease-causing organisms. Teratogenic substances can affect an unborn fetus; they may cause birth defects or spontaneous abortions or otherwise damage the fetus. Mutagenic substances cause genetic changes or mutations, which then appear in future generations.

From both an environmental and human health standpoint, persistent organic pollutants (POPs) pose the gravest threat. POPs are highly toxic, highly stable (and thus able to resist the natural processes of biodegradation), soluble in fats and oils (and therefore capable of accumulating in the fatty tissues of living organisms and biomagnifying throughout the food chain), and capable of traveling long distances in air masses. An acute exposure to high levels of POPs may be fatal or cause serious illness or injury. Chronic exposures to lower levels of these substances, however, may also be very harmful. Problems linked to long-term exposure to low levels of POPs include cancer, suppression of the immune system, and disruption of the endocrine system. Many POPs are suspected endocrine disrupters, synthetic compounds that interfere with the endocrine systems of living organisms, causing problems in growth, development, and reproduction.

Worldwide, most toxic substances are produced and used in the developed world. Pesticides are an exception; many of the most dangerous pesticides are banned for use in industrialized countries but they continue to be produced there and then exported for use in developing nations. The chemical, primary metals, and petroleum-refining industries generate and dispose of most of the world's hazardous waste. Even so, toxic chemicals can be found in garages and kitchen sinks across the United States and elsewhere in the developed world. They are considered to be household hazardous items if there are hazards involved in their use or disposal. In the United States, Congress has exempted household hazardous wastes from RCRA regulations. Thus, these toxins can be legally hauled off to a local municipal landfill or solid waste incinerator for disposal.

Just as toxic and hazardous substances adversely affect living organisms and the environment, they also exact a toll on human communities. Most industries, manufacturing plants, and landfills are located in poor communities, often inhabited by people of color, where residents have historically wielded little or no political clout. The result has been widespread environmental racism, racial discrimination in environmental policy making, enforcement of regulations and laws, and targeting of communities of color (including the lands of indigenous peoples) for the disposal of toxic wastes and the siting of polluting industries.

Hazardous wastes are disposed of primarily by incineration, landfilling, or injection underground. Incineration consists of burning organic materials at high temperatures to break them down into their constituent elements, which combine with oxygen and are released into the atmosphere. Hazardous waste landfills are designed to prevent the escape of toxic and hazardous substances into the surrounding environment. Deep-well injection involves pumping liquid wastes deep underground into rock formations. Before they are disposed of, most toxic and hazardous materials can be treated to reduce their hazardous nature. Chemical treatment processes alter the chemical structure of the constituents to produce a residue that is less hazardous than the original waste. Physical treatment, which does not affect the chemical makeup of the wastes, can take many forms, including phase separation, component separation, and solidification. Biological treatment, or bioremediation, uses microorganisms to decompose organic chemical wastes into harmless by-products.

Pollution prevention means reducing or eliminating the creation of pollutants through increased efficiency in the use of raw materials, energy, water, and other resources. Pollution prevention strategies can be grouped into two broad categories. Source reduction strategies reduce the amount of hazardous substances used and the volume of wastes produced. Source reduction was declared to be the national policy of the United States in the Pollution Prevention Act of 1990. Waste minimization strategies reduce the volume of hazardous wastes that must be disposed of. A waste exchange locates and brings together companies that have waste with companies that want to recover and reuse these resources. Pollution prevention views toxic and hazardous raw materials and wastes as resources that can and should be managed in environmentally sound ways. While it cannot eliminate all hazardous wastes, pollution prevention can enable business, industry, and households to mimic natural systems, thereby becoming less wasteful and more efficient.

In the United States, toxic and hazardous substances are managed according to the provisions of three major laws. The Comprehensive Environmental Response, Compensation, and Liability Act of 1980 (CERCLA), or Superfund, authorizes the federal government to respond directly to uncontrolled releases or threatened releases of hazardous substances into the environment. The Resource Conservation and Recovery Act of 1976 (RCRA) defines hazardous wastes; sets guidelines for managing, storing, and disposing of hazardous wastes; and establishes a cradle-to-grave regulation system, in which manifests must be kept to track wastes from generation to disposal. The Toxic

Substances Control Act (TSCA) regulates the production, distribution, and use of toxic commercial products that may present an unreasonable risk of injury to human health or the environment.

 Key Terms

acute toxicity

carcinogenic

component separation processes

endocrine disrupters

hazardous waste

infectious

mutagenic

phase separation processes

secure hazardous waste landfill

source reduction strategies

teratogenic

waste minimization strategies

bioremediation

chronic toxicity

deep-well injection

environmental racism

incineration

leachate

persistent organic pollutants

pollution prevention

solidification

still bottoms

toxic and hazardous substances

Environmental Success Story

In 1984, concerned residents in Tucson, Arizona, asked the city to help them rid their homes of hazardous waste such as paint, pesticides, and solvents to protect children from accidental exposures. The city responded quickly, providing collection facilities and dumping the offending items into a landfill. But as the cost of disposal rose, program coordinators decided to develop a way to reuse or recycle the materials. Working with the county, they established a program through which volunteers collect hazardous materials and then ship them to the county's Department of Environmental Quality (DEQ), which reuses or recycles 97 percent of them. Muriatic acid, for example, is repackaged and sent to the Recreation Department for cleaning swimming pools, and paint is repackaged and made available free of charge to local charities and low-income individuals. The program has kept 2.5 million pounds of hazardous materials out of local landfills and annual disposal costs have declined from $200,000 to $50,000, while the annual amount of materials processed has tripled.

> **Source:** Pima County Department of Environmental Quality. "Eighth Annual National Awards for Environmental Sustainability Winners: City of Tucson/Pima County Household Hazardous Waste." *Renew America* (22 April 1998). Available: http://solstice.crest.org/sustainable/renew_america /winner98.html. 24 March 1999.

True/False

1. The U.S. produces 270 million tons of hazardous waste each year.   T   F

2. All known carcinogens are also mutagens.   T   F

3. The production of toxic substances is spread evenly throughout all the regions of the United States.   T   F

4. Deep-well injection of hazardous wastes is the final solution to the disposal of nonrecyclable hazardous liquids.   T   F

5. Under the Resource Conservation and Recovery Act of 1976, the EPA is responsible for determining which wastes are considered legally hazardous.   T   F

Fill In the Blank

1. A substance which causes birth defects is known as a _____ substance.

2. Worldwide, most hazardous wastes are generated and disposed of by these three industries: _____, _____, and _____.

3. The process of entombing hazardous wastes into concrete blocks is called _____.

4. _____ are engineered to prevent the escape of leachate.

5. The Comprehensive, Environmental Response, Compensation, and Liability Act is better known as _____.

Multiple Choice

Choose the best answer.

1. Lead poisoning is an example of

    A. acute toxicity.

    B. chronic toxicity.

    C. mutagenic effect.

    D. None of the above is true.

2. All of the following are drawbacks of incinerating hazardous substances except

    A. the resultant incinerator ash can be extremely toxic.

    B. hazardous combustion products may form.

    C. particulate metals are not destroyed.

    D. the volume of disposed waste is reduced.

3. The legislation that regulates the production, distribution, and use of toxic substances is the

   A. RCRA.

   B. CERCLA.

   C. TSCA.

   D. EPA.

4. _____ is the delayed appearance of symptoms until a substance accumulates to a threshold level in the body after repeated exposures to the substance.

   A. Acute toxicity

   B. Chronic toxicity

   C. Abrupt toxicity

   D. Absolute toxicity

Short Answer

1. Explain the difference between a toxic substance and a hazardous substance.
2. What are two broad pollution prevention strategies?
3. What techniques are used in secure landfills to prevent rainwater from percolating through the wastes?
4. What are waste exchanges?
5. What are the four hazardous waste characteristics define by RCRA?

Thought Questions

Develop a complete answer for each of the following.

1. Why are the poor often at the greatest risk of exposure to toxic substances? Discuss the social issues surrounding this phenomenon.
2. Compare incineration, landfilling, and deep-well injection as methods of disposing of hazardous and toxic wastes. What are the major strengths and weaknesses of each method?
3. How does bioremediation differ from the chemical and physical methods of treating hazardous wastes?
4. Why has waste minimization become so important? Discuss strategies for reducing wastes.

5. Explain how the three major pieces of legislation are used to manage toxic substances from production to disposal. How effective is this legislation at protecting both humans and the environment from the effects of toxic substances?

▣ Related Concepts

Describe the relationship. (There may be more than one.)

BETWEEN...	AND...
acute toxicity	chronic toxicity
bottom ash	still bottoms
TSCA	RCRA
CERCLA	Superfund

Did You Know...?

The EPA has found more than 300 toxic substances in commercial oil and latex paints.

Suggested Activities

1. Start a local household hazardous materials collection program. (See *What You Can Do: To Reduce Your Exposure to Toxic and Hazardous Substances*, pages 438-439, in the textbook.)

2. Research the history of a Superfund National Priority List site. Discover how it is being cleaned up.

3. Perform a hazardous and toxic substances audit of your home. What substances do you keep around that are potentially harmful to you or the environment? Find out how they should be disposed of.

4. Begin using biodegradable and environmentally friendly substances as replacements for hazardous ones.

Chapter 21
Unrealized Resources: Waste Minimization and Resource Recovery

 **Chapter Outline**

I. **Describing Unrealized Resources: Physical Characteristics**

   A. What Is Solid Waste?

   1. Solid waste refers to any of a variety of materials that are rejected or discarded as being spent, useless, worthless, or in excess.

   2. Solid waste stream refers to the collective and continual production of all refuse.

   B. What Materials Can Be Recovered from the Solid Waste Stream?

   1. Aluminum
   2. Paper
   3. Cardboard
   4. Glass
   5. Plastics
   6. Iron and Steel
   7. Tires
   8. Used Oil
   9. White Goods
   10. Food and Yard Wastes

II. **Describing Unrealized Resources: Biological Characteristics**

   A. How Do Living Systems Manage Waste Products?

   1. Aerobic Decomposition
   2. Anaerobic Decomposition
   3. Biodegradation

   B. How Can Humans Mimic the Action of Living Systems?

   1. Natural systems are cyclic; human systems tend to be linear.

III. **Describing Unrealized Resources: Social Characteristics**

A. How Do Waste Production and Management Vary Worldwide?

1. The waste stream in LDCs is typically much smaller than that in MDCs.

B. What Misconceptions Are Associated with Solid Waste Disposal?

1. Whether by weight or volume, diapers, fast-food packaging and other plastic items are far less a problem than are newspapers, old telephone books, and yard wastes — items that easily can be recycled.

IV. **Managing Unrealized Resources**

A. What Is Ocean Dumping?

1. Problems Associated with Ocean Dumping

B. What Is Landfilling?

1. How Landfills Operate

a. Secure landfills

b. Sanitary landfills

2. Problems Associated with Landfills

C. What Is Incineration?

1. Waste-to-Energy Incinerators

a. Mass burn incinerators

b. Refuse-derived fuel incinerators

2. Problems Associated with Incineration

D. What Is Waste Minimization (or Source Reduction)?

1. An umbrella term referring to industrial practices that minimize the volume of products, minimize packaging, extend the useful life of products, and minimize the amount of toxic substances in products

E. What Is Resource Recovery?

1. Recycling

2. Municipal Solid Waste Composting

F. What Economic Factors Affect Recycling?

G. What Is a Bottle Bill?

1. Legislation that requires the consumer to pay a deposit on beverage containers which is refunded when the container is returned

H. What Is an Integrated Solid Waste System?

1. Comprehensive programs that include waste minimization, reuse of materials, recycling, composting, energy recovery, incineration of combustible materials, and landfilling

I. What Is Green Marketing?

1. The practice of promoting products based on the claim that they help or are benign to the environment

Learning Objectives

After learning the material in Chapter 21, you should able to:

1. Explain what is meant by unrealized resources and give some examples of these resources.
2. Describe how the generation and management of solid wastes vary worldwide.
3. Identify the major waste disposal methods in the United States and describe each of them.
4. Define waste minimization and resource recovery and explain why they are preferable to waste disposal.

Did You Know...?

Every day, American families produce an estimated four million pounds of household hazardous wastes (such as nail polish, paint thinner and batteries).

 ## Key Concepts

Read this summary of Chapter 21 and identify the important concepts discussed in the chapter.

A solid waste is any material that is rejected or discarded as being spent, useless, worthless, or in excess. The solid waste stream is the collective and continual production of all refuse. The two largest sources of solid waste are agriculture and mining. Municipal solid waste (MSW) is refuse generated from households, businesses, and institutions. Like hazardous wastes, solid wastes are misplaced, or unrealized, resources.

Natural systems are cyclic; in contrast, human systems tend to be linear. As the law of the conservation of matter indicates, materials can be neither created nor destroyed. The most promising solution to the solid waste problem is to reuse and recycle wastes as much as possible. Materials that can be recovered from the solid waste stream include aluminum, paper, cardboard, glass, plastics, iron and steel, tires, used oil, white goods, and food/yard wastes. For materials that cannot be recycled, the solution is to reduce consumption or to substitute nonhazardous alternatives for them.

Both the consumption of goods and the management of waste products vary from country to country. The waste stream of less-developed countries (LDCs) is typically much smaller than that of more-developed countries (MDCs).

Historically, most civilizations have managed their refuse in four ways: dumping it, burning it, converting it to something that can be used again, or reducing the amount of wastes produced.

Landfills are created by dumping refuse and then burying it beneath layers of earth and additional garbage. The Resource Conservation and Recovery Act (RCRA) established classifications for various types of landfills. Secure landfills, which have a heavy plastic liner and a clay cap, are designed to contain hazardous wastes. Sanitary landfills, lined with thick layers of clay or plastic, are built to receive nonhazardous residential, commercial, and industrial wastes. The environmental, economic, and social problems associated with landfills include closing existing landfills, rising economic costs, lack of appropriate sites for new landfills, community opposition to siting new landfills, and the possibility of leachate contamination.

As an alternative to landfilling, incineration offers some attractive benefits: it can reduce the volume of waste by 80 to 90 percent, and waste-to-energy incinerators burn garbage to produce heat and steam to generate electricity. However, air pollution and the need to landfill the remaining ash are two primary environmental problems that result.

Waste minimization, or source reduction, includes minimizing the volume of products, minimizing packaging, extending the useful life of products, and minimizing the amount of toxic substances in products. As consumers, we can all practice waste minimization via precycling. Precycling is a conscious effort to purchase merchandise that has a minimal adverse effect on the environment.

Resource recovery is an umbrella term that refers to taking useful materials or energy out of the waste stream before ultimate disposal. Recycling (collecting, processing, and marketing waste material in new products) and composting (converting organic wastes to useful soil material) are two means of resource recovery.

Interest in waste minimization and resource recovery is increasing as communities seek new ways of managing their growing waste streams. Green marketing — the practice of promoting products based on the claims that they help or are benign to the environment — is becoming more prevalent as American consumers increase their awareness of environmental issues.

Key Terms

aerobic decomposition	anaerobic decomposition
bottle bill	co-composting
co-mingled recycling	curbside collection
garbage	green marketing
integrated solid waste system	mass burn incinerator
municipal solid waste	post-consumer waste
pre-consumer waste	precycling
recycling	refuse
refuse-derived fuel incinerator	resource recovery
sanitary landfill	secure landfill
solid waste	solid waste stream
source reduction	source separation
tipping fee	trash
unrealized resources	waste minimization
waste-to-energy incinerator	

Environmental Success Story

US West Dex's Outdated Telephone Directory Recycling program illustrates how a "closed-looped" recycling program can take a bite out of the solid waste stream. Every year, this regional telephone company collects hundreds of thousands of used telephone books at drop-off sites in 14 states and recycles them into paper for the next year's directories. Since 1990, Dex has recycled more than three million used directories, saving 1.5 million trees and 610 million gallons of water. To expand the program's success, Dex introduced a school recycling program, featuring a guide book that teaches students about the benefits of recycling. One school collected more than 205 tons of old books — a 150 percent improvement over what it had collected the previous year.

Source: US West Dex. "Eighth Annual National Awards for Environmental Sustainability Winners: Outdated Telephone Directory Recycling." *Renew America* (22 April 1998). Available: http://solstice.crest.org/sustainable/renew_america/winner98.html. 24 March 1999.

True/False

1. The lowest percentage of MSW by weight is paper and cardboard. T F

2. Most of the decomposition in a landfill is aerobic. T F

3. All plastics are nonbiodegradable. T F

4. With five percent of the world's population, the United States generates about 40 percent of the world's waste. T F

5. The lifespan of a typical landfill is about 30 years. T F

Fill In The Blank

1. The two largest sources of solid waste are _____ and _____.

2. Over _____ percent of all aluminum cans are recycled, most of which are recycled into new aluminum cans.

3. A(n) _____ is charged by landfills for waste disposal.

4. Refuse may be compacted into pellets for use in a(n) _____.

5. The effort of consumers to purchase merchandise that has a minimal adverse effect on the environment is called _____.

Multiple Choice

Choose the best answer.

1. About ____ percent of the U.S. solid waste stream is municipal solid waste.

    A. 3

    B. 9

    C. 18

    D. 41

2. About 10 percent of the solid waste stream is generated by ____ activities.

    A. residential

    B. industrial

    C. agricultural

    D. mining

3. Which of the following is recycled for use in automobiles and as home insulation?

   A. aluminum

   B. glass

   C. newsprint

   D. plastic

4. Plastics account for about _____ percent of the average landfill's contents.

   A. 5

   B. 16

   C. 20

   D. 31

5. _____ landfills are designed to receive nonhazardous residential, commercial, and industrial wastes.

   A. Safety

   B. Secure

   C. Sanitary

   D. Sealed

Short Answer

1. What is the difference between a secure landfill and a sanitary landfill?
2. What are the three types of plastics that are most commonly recycled?
3. Name and describe the function of the three types of soil-dwelling bacteria that are responsible for most of the decomposition that occurs in landfills.
4. What is a "biodegradable plastic"?
5. What is green marketing?

Thought Questions

Develop a complete answer for each of the following.

1. Which elements of the municipal solid waste stream are being recycled in the U.S.? Which are not? Discuss the use of recycled materials.
2. Compare natural and human systems for handling wastes.
3. Discuss how waste production and management vary worldwide.
4. Discuss the environmental, economic, and social problems associated with landfills.

5. Discuss the economics of recycling.

■ Related Concepts

Describe the relationship. (There may be more than one.)

BETWEEN...	AND...
glasphalt	fiberglass
white goods	hazardous waste
landfill	composting
plastics	ocean dumping
waste minimization	resource recovery

Did You Know...?

About 253 million tires are thrown away every year. In Los Angeles County (California) alone, 250 tons of old tires are disposed of each day.

Suggested Activities

1. By practicing recycling, reuse, and source reduction of materials, how much can you reduce the amount of trash you throw away each week? Start your own integrated waste management program. Keep a record of the types of materials and how you handle them. If you already do some of these things, see if you can increase the amount you recover or reduce.

2. Where does the trash generated in your community go? Investigate and report on your findings.

3. Reduce the amount of waste you produce by precycling. (See *What You Can Do: To Precycle*, page 462, in the textbook.)

4. Support deposit legislation. Urge your state representatives to pass a statewide bottle bill, and ask your members of Congress to support deposit legislation nationwide.

Student Learning Guide Chapter 21 **193**

UNIT VI

An Environmental Heritage:
Preserving Threatened Resources

Chapter 22
The Public Lands

 Chapter Outline

I. **Describing the Public Lands Resource: Biological Characteristics**

  A. What Are the Federal Public Lands?

   1. The federal government manages most public lands, an area equivalent to approximately one-third of the entire United States.

   2. More than 90 percent of public lands are under the jurisdiction of just four agencies:

     a. National Park System

     b. National Wildlife Refuge System

     c. National Forest System

     d. National Resource Lands

  B. Why Are the Public Lands Biologically Significant?

II. **Describing the Public Lands Resource: Physical Characteristics**

  A. Mineral Wealth

  B. Renewable Resources

  C. Electricity from Steam and Hot Water

III. **Describing the Public Lands Resource: Social Characteristics**

  A. Consumptive Uses

   1. Use up some portion of the resource in the process (such as logging, livestock grazing, and mining)

  B. Nonconsumptive Uses

   2. Do not deplete the resource (such as bird watching, photography, and nature study)

IV. **Managing the Public Lands**

  A. How Did the System of Federal Lands Develop?

  B. How Did the National Park System Develop?

  C. What Environmental Problems Face the National Park System?

   1. Internal Threats to the National Park System

        a. Overuse

        b. Insufficient Funding and Park Operations

        c. Threats to Wildlife

        d. Concessions System

        e. Energy and Minerals Development

   2. External Threats to the National Park System

        a. Atmospheric Pollution

        b. Activities on Adjacent Lands

D. How Did the National Wildlife Refuge System Develop?

E. What Environmental Problems Face the National Wildlife Refuge System?

   1. Internal Threats to the National Wildlife Refuge System

        a. Management Structure

        b. Secondary Uses

   2. External Threats to the National Wildlife Refuge System

        a. Activities on Adjacent Lands

        b. Political Pressures

F. How Did the National Forest System Develop?

   1. Utilitarianism — Conservation For Economic Reasons

G. What Environmental Problems Face the National Forest System?

   1. Emphasis on Logging Over Other Uses

   2. Reliance on Clearcutting

   3. Below-Cost Timber Sales

   4. Reluctance to Designate Wilderness Areas

H. How Did the Network of National Resource Lands Develop?

I. What Environmental Problems Face the National Resource Lands?

   1. Illegal Harvesting and Disruptive Recreational Use

   2. Commercial Exploitation

   3. Reluctance to Designate Wilderness Areas

Learning Objectives

After learning the material in Chapter 22, you should be able to:

1. Discuss the biological, physical, and social significance of the public lands resource.
2. Identify and briefly describe the four major types of federal lands and the agencies that manage each.
3. Briefly explain the historical development of the system of federal public lands.
4. Describe the major environmental problems and management issues associated with the National Park System, National Wildlife Refuge System, National Forest System, and National Resource Lands.

Did You Know...?

America loses about 300,000 to 400,000 acres of wetlands each year.

 ## Key Concepts

Read this summary of Chapter 22 and identify the important concepts discussed in the chapter.

The public domain, or land holdings of the federal government, is so vast that as much as one-half of it has never been surveyed. Currently, it includes approximately 726 million acres of forest, desert, grassland, wetland, and other lands. They are managed by numerous federal agencies, chiefly the Park Service (Department of Interior), Fish and Wildlife Service (Department of Interior), Forest Service (Department of Agriculture), and Bureau of Land Management (Department of Interior).

The biological significance of the federal lands has to do with the variety of ecosystems they contain, the biological diversity they harbor, and the ecosystem services they perform. The physical boundaries of the federal lands pertain to their mineral wealth; these areas contain one-quarter of the nation's coal, four-fifths of its huge oil-shale deposits, one-half of its uranium deposits and naturally occurring steam and hot water pools, one-half of its estimated oil and gas reserves, and significant reserves of strategic minerals.

The social significance of the federal lands has to do with how they are used. Commercial consumptive uses include logging, mining, and oil development. Nonconsumptive recreational uses include scientific study, wilderness travel, hiking, bird watching, and canoeing. It is important to realize that nonconsumptive uses such as hiking or trail biking can degrade an area if there are many users or if the users are not careful about the way in which they use the land.

The history of the federal lands is largely one of land disposal to private interests. This tradition dates back to the early part of the nineteenth century when land disposal was seen as a way of encouraging western expansion and settlement and thus securing the nation's frontier. By the mid-1930s, the government had either given away or sold 1.1 billion acres.

Despite its tradition of land disposal, the government had begun to set aside land units for special purposes as early as 1832. Forty years later, Yellowstone was designated a "nation's park." The 1891 Forest Reserve Act facilitated the establishment of the national forests; this law was also invoked to establish the first national wildlife refuge in 1903. By the 1930s, the majority of federal lands remained unreserved, not belonging to the National Park System, the National Wildlife Refuge System, or the National Forest System. The public domain, as it was called, was severely degraded due to overgrazing and other deleterious practices. Finally, in the late 1940s, the Bureau of Land Management (BLM) was formed to oversee the public domain.

The National Park System encompasses 378 diverse units totaling over 83 million acres. The Park Service strives to manage the parks in order to conserve their scenery, natural and historic objects, and wildlife, and to provide for the enjoyment and use of the same by the public. Major threats to the Park System are crowding and overuse, insufficient funding and park operations, diminishing habitat, elimination of predators, poaching, commercial recreational opportunities, energy and minerals development, atmospheric pollution, and activities conducted on adjacent lands.

The National Wildlife Refuge System (NWRS) is a network of 550 units totaling about 93 million acres of land and water. The NWRS provides habitat and haven for wildlife. Major threats to the National Wildlife Refuges are the internal management structure of the U.S. Fish and Wildlife Service (USFWS), which is plagued by too many competing responsibilities; harmful or incompatible secondary uses of many refuges; activities on adjacent lands; and political pressures.

The National Forest System encompasses approximately 191 million acres in 44 states, Puerto Rico and the Virgin Islands. The nation's forests and grasslands are multiple use lands — they accommodate both commercial and recreational uses as well as wilderness habitat. Major threats to the National Forests are an emphasis on logging over other uses, clearcutting, below cost timber sales, and lack of wilderness designation/protection.

Multiple use is also the philosophy that governs the National Resource Lands, or Bureau of Land Management (BLM) lands. Found chiefly in the arid and semi-arid western states and Alaska, BLM lands equal approximately 264 million acres, almost half of all federal land. Even though they form the largest block of federal lands, BLM lands receive relatively few visitors. Major threats to the National Resource Lands include looting, poaching and illegal harvesting, commercial exploitation (especially grazing), and lack of wilderness designation.

 **Key Terms**

below-cost timber sales

consumptive use

multiple use

public domain

uneven-aged harvesting

clearcutting

even-aged harvesting

nonconsumptive use

shelterwood

utilitarianism

Environmental Success Story

Located in Nelsonville, Ohio, Hocking College is demonstrating that restoring the environment and improving the economy can go hand-in-hand. The College's National Environmental Training Cooperative (NETC) binds together labor and management, government and private and public organizations. Together they educate the public; restore environmentally unstable land, water, and air; and retrain displaced workers for emerging private sector environmental jobs. Focusing on job losses in the local mining industry, NETC has targeted coal miners first for retraining. To date, 17 miners have been trained in environmental restoration; 14 of these employees have full-time jobs with benefits. In the training process, 20 acres of abandoned mining lands have been restored — sites which might not otherwise have been remediated. Now these lands, which NETC restored at a reduced cost, are used for farming, wildlife areas, and parks.

Source: National Environmental Training Cooperative. "Seventh Annual National Awards for Environmental Sustainability Winners: Hocking College School of Natural Resources and Ecological Sciences." *Renew America* (22 April 1998). Available: http://solstice.crest.org/environment/renew_america/winners.html. 24 March 1999.

True/False

1. Nonconsumptive uses are not considered a threat to the National Parks.   T   F

2. The public lands account for approximately one-third of the total area of the U.S.   T   F

3. Mining is permitted in National Forests.   T   F

4. BLM lands are located only in Alaska and the western states.   T   F

5. The Department of the Interior and the Department of Agriculture share the responsibility for all public lands.   T   F

Fill in the Blank

1. The BLM lands are officially known as _____.

2. Bird watching and photography are examples of _____.

3. President Theodore Roosevelt broadened the intent of the _____ to preserve significant natural areas.

4. Historically, the _____ were set aside to protect places of spectacular scenery or natural grandeur.

5. _____ occur when the price paid by loggers for timber is lower than the costs incurred by the Forest Service to make the timber available.

Multiple Choice

Choose the best answer.

1. Which has the largest land area?

    A. National Park System

    B. National Wildlife Refuge System

    C. National Forest System

    D. National Resource Lands

2. The National Forest System is managed by the

    A. National Park Service.

    B. U.S. Fish and Wildlife Service.

    C. Department of Agriculture.

    D. Department of the Interior.

3. Which agency manages the National Wildlife Refuge System?

    A. National Park Service

    B. U.S. Fish and Wildlife Service

    C. Department of Agriculture

    D. None of the above is true.

4. The purpose of the National Forest System is to

    A. provide habitat for wildlife.

    B. manage for sustained multiple use.

    C. preserve the natural scenery and provide for use by the public.

D. None of the above is true.

5. Historically, which practice governed the management of the public domain?

    A. multiple use

    B. disposal to private interests

    C. protectionism

    D. isolationism

Short Answer

1. What are some of the problems plaguing wildlife refuges?
2. What is a split estate?
3. What is a duck stamp?
4. List the five land uses the National Forest Service is mandated to balance.
5. Name the legislation that formed the National Park Service.

Thought Questions

Develop a complete answer for each of the following.

1. Explain the biological significance of the public lands.
2. Discuss the conflicts between consumptive and nonconsumptive uses of the federal lands.
3. Describe the major threats to the National Parks.
4. How does the structure of the Fish and Wildlife Service threaten the National Wildlife Refuge System?
5. How does the concept of utilitarianism influence the management of National Resource Lands and National Forests? Why would the managing agencies sell grazing rights and timber at below cost?
6. Discuss the effects of having the four types of federal lands managed by different agencies. How do conflicting interests strive to determine land use policy?
7. Explain the reasons behind clear-cutting and describe the impact on the environment.

▣ Related Concepts

Describe the relationship. (There may be more than one.)

BETWEEN...	AND...
clear-cutting	flash flooding
fire management policy in Yellowstone National Park	lodgepole pines
commercial use	National Resource Lands
concessions	National Parks

Did You Know...?

Below-cost timber sales resulted in average annual losses by the Forest Service of $327 million between 1979 and 1990.

 ## Suggested Activities

1. Pick a public land area that interests you and get to know its natural and cultural history.

2. Join a conservation organization to support its efforts to protect public lands.

3. Visit a national park, forest, refuge or BLM land. Practice low-impact camping and hiking.

4. Let your elected officials know where you stand on issues pertaining to public lands.

Chapter 23
Wilderness

 Chapter Outline

I. **Describing the Wilderness Resource: Biological Characteristics**

   A. What Is Wilderness?

   1. Wilderness ecosystem — an ecosystem in which both the biotic and abiotic components are minimally disturbed by humans

   B. How Do Wilderness Areas Preserve Biological Diversity?

   1. Genetic Diversity

   2. Species Diversity

   3. Ecosystem Diversity

II. **Describing the Wilderness Resource: Physical Characteristics**

   A. Most protected wilderness are high alpine and tundra ecosystems, and most are in the western states and Alaska.

III. **Describing the Wilderness Resource: Social Characteristics**

   A. What Is the Cultural Significance of Wilderness?

   1. Cultural Heritage

   B. Why Should We Preserve Wilderness?

   1. Ecological Importance

   2. Personal Benefits

   3. Economic Benefits

IV. **Managing Wilderness**

   A. How Did the Wilderness Preservation Movement Develop?

   1. Arthur Carhart

   2. Aldo Leopold

   3. Robert Marshall

   B. How Were Wilderness Areas Managed Historically?

   1. Utilitarian philosophy — emphasized economic uses over recreation and wilderness preservation

Student Learning Guide Chapter 23

    C. What Is the Wilderness Act of 1964?

        1. Howard Zahniser

        2. Hubert H. Humphrey

    D. How Is an Area Designated as Wilderness?

    E. How Did the Wilderness Act Affect the Management of the Federal Lands?

    F. What Is Roadless Area Review and Evaluation (RARE)?

    G. What Is the Federal Land Policy and Management Act?

    H. How Are Wilderness Areas Currently Managed?

    I. What Environmental Problems Face Wilderness Areas?

        1. Pressures Caused by Heavy Recreational Use

        2. Lack of Wilderness Designation

    J. How Can We Improve Wilderness Management?

Learning Objectives

After learning the material in Chapter 23, you should be able to:

1. Define "wilderness" and identify the single characteristic common to all definitions of wilderness.

2. Discuss the reasons most often given for preserving wilderness.

3. Summarize the cultural roots of the wilderness preservation movement and the development of the Wilderness Preservation Act of 1964.

4. Explain the process for designating an area as wilderness.

5. Briefly relate how wilderness areas are managed.

6. Identify the major threats to wilderness areas.

> **Did You Know...?**
>
> At 9,078,675 acres (3,674,089 hectares), Alaska's Wrangell-St. Elias National Park is the largest protected wilderness area in the United States.

Key Concepts

Read this summary of Chapter 23 and identify the important concepts discussed in the chapter.

A wilderness ecosystem is one in which both the biota and abiota are minimally disturbed by humans. Not surprisingly, then, wilderness areas serve as reservoirs of biological diversity, or biodiversity. The single best hope to preserve maximum biodiversity is to preserve habitats (ecosystems) as wilderness. The United States's National Wilderness Preservation System, a network of some 445 units covering over 89 million acres, protects just 81 of the country's 233 distinct ecosystem types. Many ecosystems, and their resident biota, remain unprotected.

Those who argue in favor of protecting wilderness point to the personal (mental and physical well-being), economic, and ecological benefits (ecosystem services) it provides, and to the fact that it is an integral part of our nation's cultural heritage. Opponents to wilderness preservation argue that it "locks up" valuable resources which could be used for economic development and that it benefits only a minority of recreational users.

Fifty years ago, the idea of protecting and preserving wilderness was virtually unheard of. After all, for centuries, humans had feared "wild" lands. (Interestingly, the root word for wilderness is the same as the root word for bewildered — to be confused.) In fact, since the advent of the European conquest of North America, settlers had sought to "tame" the wilderness and transform the landscape into the pastoral vistas they had left behind in Europe. A few lone voices — Henry David Thoreau and John Muir, most notably — sung the praises of wild lands and warned of the threat posed by unchecked development, but most people did not heed their warnings. By the early 1930s, others were sounding the alarm. Aldo Leopold, Arthur Carhart, Robert Marshall, Benton MacKaye, and Robert Sterling Yard were some of the earliest proponents of a national system of protected wilderness areas. Unfortunately, with the building boom that followed World War II, economic uses of land were emphasized over recreation and wilderness preservation. A long struggle ensued which pitted wilderness preservationists against developers and other economic forces. Finally, in 1964, the Wilderness Preservation Act was passed to counteract the emphasis on development and economic uses of the land. The Act represented the culmination of years of effort by many people, notably Howard Zahniser, who authored the legislation but did not live to see it enacted. The Wilderness Preservation Act designated nine million acres of federal land as protected wilderness. Though the Forest Service developed regulations for the protection and management of areas already classified as wilderness, it did not recommend any additions to the system between 1964 and 1973. To correct this situation, Congress passed the Wilderness Act of 1973. By the late 1980s, the wilderness system had grown ten-fold, to some 89 million acres.

Only lands included in the National Wilderness Preservation System are officially managed as protected wilderness. Both officially designated areas and de facto wilderness areas face numerous threats to their ecological integrity. Among the threats to de facto wilderness are the vagaries of the designation process itself, chaining, overgrazing, and the lack of federal water rights. Threats to designated areas include increased crowds, litter, vandalism (especially at archeological sites), damaged vegetation, water pollution, and soil erosion from heavily used trails and campsites.

Key Terms

chaining

reserved water rights

de facto wilderness

wilderness ecosystem

Environmental Success Story

When Howard Hughes, owner of the Ballona Wetlands, died in 1976, his heirs intended to develop the land. In response, a group of concerned citizens formed the Friends of Ballona Wetlands (FBW), an organization dedicated to preserving the last remaining wetlands in Los Angeles. Through aggressive lobbying, public education, and, finally, a lawsuit, FBW not only prevented massive development on the site, but also secured $12 million from the landowner to restore existing damage. In 1997, more than 3,000 volunteers of all ages helped plant over 65 native plant species and 25 native trees. FBW recruits children from low-income and minority families in the Los Angeles area to help plant native species and remove nonnative ones, which assists in the area's recovery and helps instill in these young people a sense of environmental ownership and pride. Once restoration is complete, the hope is for the wetlands to once again serve as a habitat for a variety of now-endangered species, including the El-Segundo blue butterfly and the Silvery legless lizard.

> **Source:** Friends of Ballona Wetlands. "Eighth Annual National Awards for Environmental Sustainability Winners: Ballona Dunes Restoration Education Program." Renew America. (22 April 1998). Available: http://solstice.crest.org/sustainable/renew_america/ winner98.html. 24 March 1999.

True/False

1. The National Wilderness Preservation System represents at least one of each type of ecosystem found in the U.S.   T   F

2. Most designated wilderness areas are found within National Parks.   T   F

3. Most protected wildernesses are high alpine or tundra ecosystems.   T   F

4. Mining or grazing is not allowed in any wilderness area.   T   F

5. The National Park Service opposed the Wilderness Act of 1964.   T   F

Fill in the Blank

1. The key legislation which protects wilderness areas is the _____.

2. In 1935, Robert Marshall and other preservationists formed the _____ to advocate for wilderness preservation.

3. An area being considered for inclusion in the National Wilderness Preservation System is known as a(n) _____.

4. Wild lands that are undesignated and unprotected are called _____.

5. A controversial BLM procedure used to reclaim rangelands by using two bulldozers and a chain to tear down trees and shrubs is called _____.

Multiple Choice

Choose the best answer.

1. How many ecosystem types are represented in the National Wilderness Preservation System?

    A. 50

    B. 81

    C. 100

    D. 233

2. As defined by the Wilderness Preservation Act of 1964, the most important characteristic of wilderness is

    A. biological diversity.

    B. biotic and abiotic components.

    C. minimal disturbance by humans.

    D. far removed from civilization.

3. Under the Wilderness Act, who has the responsibility for designating wilderness areas?

    A. Congress

    B. National Park Service

    C. National Forest Service

    D. All of the above are true.

4. The biggest threat to undesignated wilderness areas is

   A. logging.

   B. mining.

   C. being left out of the wilderness system.

   D. chaining.

5. The fight over which national treasure led to the development of the Wilderness Preservation Act of 1964?

   A. Yellowstone

   B. Yosemite

   C. Boundary Waters Canoe Area

   D. Dinosaur National Monument

Short Answer

1. What, according to early preservationist Robert Marshall, are the most important attributes of a wilderness area?

2. Who was Howard Zahniser?

3. List three of the five objectives of wilderness management under the 1964 Wilderness Act.

4. What was RARE?

5. What are the ecological benefits of wilderness preservation?

Thought Questions

Develop a complete answer for each of the following.

1. Discuss the arguments against wilderness designation and the corresponding defense made by conservationists.

2. Discuss the value of wilderness and explain the nature of the arguments for preserving wilderness.

3. How did the wilderness preservation movement develop in the 1920s and 1930s? What were the major obstacles faced by early preservationists?

4. How are wilderness areas managed?

5. Describe the environmental problems currently facing designated wilderness areas and explain what is being done to protect them.

▢▨▪ Related Concepts

Describe the relationship. (There may be more than one.)

BETWEEN...	AND...
Arthur Carhart	Aldo Leopold
utilitarianism	wilderness preservation
Forest Service	Wilderness Act of 1973
1976 Federal Land Policy and Management Act	Wilderness Act of 1964
Bureau of Land Management	wilderness designation

Did You Know...?

Scientists have identified 233 distinct ecosystem types in the United States, but only 81 are represented in the wilderness system.

 ## Suggested Activities

1. Choose one of the members of the wilderness preservation movement mentioned in this chapter or one you discover on your own, and find out more about him or her and what he or she did in the fight to preserve wilderness.

2. Visit a wilderness area. Document your experience.

3. Locate a wilderness area on a map. Research it to discover its unique qualities.

Chapter 24
Cultural Resources

 Chapter Outline

I. **Describing Cultural Resources: Physical and Biological Characteristics**

   A. What Are Cultural Resources?

   1. Material Culture

      a. Tangible objects, such as tools, furnishings, buildings, sculpture, and paintings

   2. Nonmaterial Culture

      a. Intangible resources, such as language, customs, tradition, and folklore

   B. Where Are Cultural Resources Found?

II. **Describing Cultural Resources: Social Characteristics**

   A. Why Are Cultural Resources Significant?

   1. Historical Record

   2. Symbols of Our Heritage

   3. Cultural Identity

   4. Storehouse of Environmental Knowledge

   5. Economic Benefits

   6. Cultural Diversity

   B. What Are the Threats to Cultural Resources?

III. **Managing Cultural Resources**

   A. How Have U.S. Cultural Resources Been Preserved?

   1. Early Government Involvement

      a. The Antiquities Act of 1906

   2. World War II and Postwar Economic Growth

      a. Urban renewal

   3. Preservation Efforts Since 1960

      a. Major legislation

      b. Urban revival

            c. Applied ethnography
    B. How Have Cultural Resources Been Preserved Worldwide?
        1. United Nations Education, Scientific, and Cultural Organization (UNESCO)

Learning Objectives

After learning the material in Chapter 24, you should be able to:

1. Define cultural resource and differentiate between material and nonmaterial culture.
2. Explain why it is important to preserve cultural resources.
3. Support the relevance of studying environmental history.
4. Provide examples of environmental problems that hinder the preservation of cultural resources.
5. Provide examples of human actions that hinder the preservation of cultural resources.
6. Discuss at least three national and three international efforts to preserve cultural resources.

Did You Know...?

Tribal women in India know of medicinal uses for some 300 forest species.

 Key Concepts

Read this summary of Chapter 24 and identify the important concepts discussed in the chapter.

A cultural resource is anything that embodies or represents a part of the culture of a specific people. All cultural resources arise from human thought or action. They often reflect or are part of the environment. Material cultural resources are tangible objects — tools, furniture, monuments, and artwork, for example — that humans create to make life easier or more enjoyable. Nonmaterial cultural resources include intangible objects such as traditions, customs, and folklore. Cultural resources management is concerned with preserving both material and living culture.

Cultural resources can be found virtually everywhere. They are significant because (1) they provide us with a record of societies and their environments, (2) are symbols of our heritage, (3) are integral to cultural identity, (4) are a valuable storehouse of environmental knowledge, (5) provide economic benefits, and (6) are essential for cultural diversity. Material cultural resources may be damaged because of the location of a structure, the ground on which it sits, the soil type, faulty materials, and building defects. They may also be damaged by long-term natural causes (such as rain, frost, humidity, vining vegetation, and animal droppings). Air pollution, the use of wells, and the construction of tunnels can also threaten material culture. Others factors which may damage material culture include changing tastes in fashion, war, development, vandalism, looting, and pillaging. The greatest threat to nonmaterial culture is acculturation, the process by which one culture adapts or is modified through contact with another.

Until fairly recently, the private sector was the force behind the preservation movement in the United States. The Antiquities Act of 1906 was the country's first major piece of legislation to safeguard cultural resources and was followed throughout the century by a host of additional acts designed to strengthen resource protection. In an attempt to preserve the cultures of indigenous peoples, the National Park Service in 1991 began an applied ethnography program, which promotes increased ethnic community involvement in the administration of several park units.

The chief international agency working to protect the global cultural heritage is UNESCO, the United Nations Educational, Scientific, and Cultural Organization. UNESCO member nations agree to a binding legal framework that obliges them to protect global cultural heritage. UNESCO can establish international conventions concerning the protection of cultural resources. Countries are free to ratify or deny a convention as they choose, but if they do ratify it, they are bound to comply with its stipulations. So far, UNESCO has enacted three conventions. One convention concerned the protection of cultural resources in case of an armed conflict; another established guidelines to prohibit and prevent the illegal import, export, and transfer of ownership of cultural resources. The third convention established the World Heritage List of objects and sites of global importance.

Key Terms

acculturation	applied ethnography
cultural resource	cultural resources management
historic preservation	material culture
nonmaterial culture	rehabilitation
restoration	urban renewal
urban revival	

Environmental Success Story

About a decade ago, at-risk youth counselor Neil Andre took his first camping trip with a group of inner-city children and discovered something: Not only were the children unfamiliar with nature but they also were afraid of it. Determined to change this situation, Andre began introducing environmental games into activities with the children. Soon the games turned into community projects, and then the projects grew so numerous that Andre and the children formed an organization dedicated solely to environmental work. Now numbering 150 children, Earth Angels is a unique organization in which at-risk youth discuss, choose, and pursue their own environmental projects. Over the past several years, Earth Angels has recycled thousands of pounds of aluminum, glass, and tires. Members have planted dozens of trees, and they have bought and preserved 42 acres of rain forest in Latin America through The Nature Conservancy. Through their work, the Earth Angels children are learning how to lead, cooperate, and develop a sense of hope about the future.

Source: Guardian Angel Settlement Association. "Eighth Annual National Awards for Environmental Sustainability Winners: Earth Angels." *Renew America* (22 April 1998). Available: http://solstice.crest.org/sustainable/renew_america/winner98.html. 24 March 1999.

True/False

1. Material culture includes such things as language, customs, traditions, and folklore.   T   F

2. Indigenous or tribal people tend to possess a broad and deep knowledge of their ecosystems.   T   F

3. The outward features of a culture can be abandoned more easily than can cultural knowledge and values.   T   F

4. Both material and nonmaterial cultural resources are threatened by extrinsic factors.   T   F

5. Until fairly recently, the federal government was responsible for most historic preservation in the United States.   T   F

Fill in the Blank

1. _____ is the field concerned with preserving material cultural resources.

2. Folklore is an example of _____ culture.

3. _____ is the process by which one culture adapts or is modified through contact with another.

4. Faulty building materials and design defects are examples of _____ threats.

5. The National Historic Preservation Act created the _____, a list of historic sites of local, state, and regional significance.

Multiple Choice

Choose the best answer.

1. Which of the following is not a threat to cultural resources?

    A. vandalism

    B. cultural diversity

    C. war

    D. air pollution

2. What is the act or process of returning a property to a state of utility (through repair or alteration) in order to accommodate an efficient contemporary use, while preserving those features or portions of the property that are historically, architecturally, or culturally significant?

    A. rehabilitation

    B. restoration

    C. urban revival

    D. remediation

3. Which of the following is (are) given as reasons to preserve cultural resources?

    A. They provide a historical record of societies and their environments.

    B. They serve as symbols of our culture.

    C. They provide a people with a cultural identity.

    D. All of the above are true.

4. Which of the following is an example of nonmaterial, or living, culture?

    A. Statue of Liberty

    B. Egyptian pyramids

    C. Native American languages

    D. Golden Rock of Burma (Mynamar)

5. How does pillaging affect a historic or cultural site?

    A. Removes important cultural resources, which are national property (not for private use)

    B. Disrupts and thereby ruins the archeological record

    C. Desecrates a site that may have religious significance and thereby degrades its spiritual quality

    D. All of the above are true.

Short Answer

1. What is the difference between historic preservation and cultural resources management?

2. Explain how acid precipitation threatens material cultural resources.

3. How does the National Park Service's applied ethnography program promote more appropriate management of park units?

4. What is UNESCO?

5. What is the World Heritage List?

Thought Questions

Develop a complete answer for each of the following.

1. Why are cultural resources important and worthy of preservation? List and explain at least five reasons. Are there any arguments for not preserving cultural resources?

2. What factors threaten both material and nonmaterial culture? List and describe at least seven.

3. What is the connection between cultural resources and the environment? Between the study of cultural resources and environmental science? Are cultural resources an appropriate topic in an environmental science text? Why or why not?

4. Briefly summarize the history of cultural resources preservation in the United States. What forces gave rise to the modern preservation movement?

5. What is being done to preserve cultural resources in the United States and worldwide? What are the major organizations and legislation involved?

▢▦■ Related Concepts

Describe the relationship. (There may be more than one.)

BETWEEN...	AND...
acculturation	applied ethnography
acid precipitation	extrinsic threat
pillaging for profit	Archeological Resources Protection Act of 1979
urban renewal	historic preservation
cultural resources	environment

Did You Know...?

Massive deforestation, combined with the spread of disease, threatens an estimated 1,000 tribes of indigenous people with extinction.

 Suggested Activities

1. Discover the past of a historic site.

2. Visit a natural history museum or historical society.

3. Learn about the customs of a culture other than your own.

4. Study the history and customs of your own family. Talk to relatives to find out where these traditions came from.

UNIT VII

An Environmental Legacy:
Shaping Human Impacts on the Biosphere

Chapter 25
Religion and Ethics

 Chapter Outline

I. **What is Religion?**

    A. Expression of the human belief in and reverence for a superhuman power

    B. Characterized by a core of beliefs which answer questions about God, the universe, and humankind

    C. Two important religious questions reveal humanity's place in world:

        1. Are humans part of, or apart from, nature?

        2. What is humanity's role in creation — to act as master or as steward?

II. **How Does Religion Affect Environmental Problem Solving and Management?**

    A. Anthropocentric Worldview — Humans dominate the natural world.

    B. Biocentric Worldview — Humans are no more or less valuable than all other parts of creation.

III. **How Are Religions Promoting Environmentally Sound Management?**

    A. Some Modern Examples

        1. North American Conference on Christianity and Ecology (NACCE)

        2. The Buddhist Peace Fellowship (BPF)

        3. The Southwest Environmental Equity Project (SWEEP)

        4. Floresta

        5. Orthodox Christianity Under the Leadership of Bartholomew I, "The Green Patriarch"

        6. The National Religious Partnership for the Environment (NRPE)

        7. Evangelical Environmental Network

        8. Earth-based Spiritualities and Traditions

IV. **What Are Ethics?**

    A. Branch of philosophy concerned with standards of conduct and moral judgment

    B. System/code which shapes attitudes and behaviors toward others and the world

V. How Do Ethics Affect Environmental Problem Solving and Management?

A. Frontier Ethic

1. According to settlers' interpretation of Christianity, resources were to be used fully in a quest to "subdue the Earth."

2. Land and natural resources were exploited without regard for ecosystem.

B. Environmental Ethic — Humans as Part of the Natural World

C. Stewardship Ethic — Humans as Caretakers of the Natural World

D. Land Ethic

1. First formal statement of an environmental ethic

2. Developed approximately fifty years ago by Aldo Leopold

3. Defined as a limitation of freedom in the struggle for existence

4. Extends ethic to include community — other living species, soils, water, or collectively, the land

5. Reaffirms the right of natural resources to exist

VI. How Can We Apply a Personal Ethic to Societal Problems?

A. Variability of personal ethics makes them difficult to apply to societal problems.

B. Examples of environmental subjects raising ethical questions:

1. California condors removed from wild for captive propagation

2. Pacific yew — slow-growing tree found in old-growth forests that contains taxol, a cancer-fighting substance

3. Global commons — especially oceans and atmosphere

Learning Objectives

After learning the material in Chapter 25, you should be able to:

1. Explain what religion is and how religious beliefs affect a person's or society's interactions with the environment.

2. Distinguish between anthropocentric and biocentric worldviews.

3. Give some examples of how religions are promoting environmentally sound management.

4. Explain what an ethic is and distinguish between an environmental ethic and a frontier ethic.

5. Identify at least two current situations that pose an ethical dilemma and explain them in terms of environmental consequences.

> **Did You Know...?**
>
> 66 percent of all Americans consider themselves environmentalists.

 ## Key Concepts

Read this summary of Chapter 25 and identify the important concepts discussed in the chapter.

Religion is the expression of human belief in, and reverence for, a superhuman power recognized as a supreme being, a supernatural realm, or an ultimate meaning. Religions are characterized by a unique core of beliefs or teachings that answer basic questions about the universe, existence, the world, and humankind. The values, beliefs, and attitudes inspired by religions affect, to varying degrees, how the faithful respond to specific environmental concerns. Religion is an important component of a people's worldview.

Religions can affect environmental problem solving and management in several ways: by preventing people from taking action on problems in the belief that God will take care of all things; limiting the actions or options of its followers in a given situation; and shaping followers' worldviews (which affect beliefs, values, attitudes, and behavior).

The past few decades have seen a shift in the thinking of some organized religions. Concern for the Earth and the belief that humans are to act as stewards of creation are increasingly widespread ideas among many religious groups.

Ethics is the branch of philosophy concerned with standards of conduct and with distinguishing behavior which is right from that which is wrong. An individual's ethic, or personal code of right and wrong behavior, is often based on his or her morals, principles that help one distinguish between good and evil. Morals imply a faith component — faith in God or faith in an ultimate, universal meaning or truth. Aldo Leopold, in *A Sand County Almanac,* maintained that ethics should be extended from their purely social context to the land and environment. He argued for the development of a land ethic to guide our behavior toward the environment. Such an ethic, he maintained, would transform humans from conqueror of the biotic community to "plain member and citizen of it." Leopold pointed out than a land ethic would imply respect for the biotic community and for the environment as a whole.

There is no universal human ethic. Moreover, ethical codes change over time; what seems correct and appropriate in one era may be viewed as immoral in another. For example, the frontier ethic, which fostered aggressive behavior toward the environment, is gradually being replaced by a stewardship ethic, in which humans are seen as caretakers of the natural world. The stewardship ethic is a type of land ethic.

It is difficult to apply ethical standards in societal situations because many people have widely different ideas about what is ethical behavior, and in many societal issues, there is no clear right and wrong. In the case of the California condor, the Pacific yew tree and the global commons, there is an on-going debate to decide how we as a society should proceed.

 Key Terms

anthropocentric worldview	biocentric worldview
environmental ethic	ethic
ethics	frontier ethic
immanence	interconnection
land ethic	morals
religion	stewardship ethic

Environmental Success Story

When the nearby sewage treatment plant was faced with too much waste to handle, the Ben & Jerry's Homemade Ice Cream plant decided to help by diverting some of its milky waste. And to where did they divert their waste? A pig farm! The company bought 250 piglets for the farmer while providing him with the solids from the milk to prepare the pigs for market. The only condition that the farmer had to agree to was naming three of his pigs Ben, Jerry, and Ed Stanek. Ed is the head of Vermont's Environmental Commission and a supporter of Ben & Jerry's innovative waste reduction method.

True/False

1. About one-third of the world's population professes belief in some form of Christianity.   T   F

2. The frontier ethic was the first formally developed statement of an environmental ethic.   T   F

3. A stewardship ethic may be based on either a biocentric or anthropocentric worldview.   T   F

4. Ethics place limits on human actions.   T   F

5. Religions do not affect how people solve environmental problems and manage resources.   T   F

Fill in the Blank

1. Aldo Leopold promoted a(n) _____ ethic in his book, *A Sand County Almanac*.

2. The belief that a divine entity is embodied in the living world is known as _____.

3. The _____ are a Protestant group that practices low-input agriculture and self-reliant living.

4. The _____ ethic encourages the exploitation of natural resources.

5. The _____ has been found to contain taxol, a cancer-fighting agent.

Multiple Choice

Choose the best answer.

1. St. Francis of Assisi has been proposed as the patron saint of

    A. ethics.

    B. religion.

    C. ecology.

    D. Italy.

2. A moral is a(n)

    A. ending to a story.

    B. principle that helps an individual distinguish right from wrong.

    C. majority, in some cases.

    D. ethical decision.

3. Which of the following is considered to be a *global* commons?

    A. coastal mangrove forests

    B. tropical rain forests

    C. ocean fisheries

    D. biological resources

4. According to Edward Abbey, the "conscience of our race" is

    A. absent.

    B. ecology.

    C. environmentalism.

    D. ethics.

5. The two core concepts central to all Earth-based spiritualities are:

   A. immanence and interconnection.

   B. stewardship and anthropocentrism.

   C. stewardship and biocentrism.

   D. feminism and rebirth.

Short Answer

1. What factors other than religious tenets play a role in determining people's behavior toward the environment?

2. What is Floresta?

3. List three traditions of Earth-based spiritualities.

4. What is an ethic?

Thought Questions

Develop a complete answer for each of the following.

1. Discuss how religious beliefs and ethical codes affect the environment.

2. How are organized religions currently handling environmental issues and problems? In what ways are they promoting environmentally sound management?

3. Do anthropocentric worldviews always result in non-sustainable systems or harmful effects on the environment? Explain your answer.

4. Discuss why it is difficult to apply personal ethical codes to social problems.

5. Explain the concept of the global commons. What are the implications in terms of managing shared resources?

▢▨■ Related Concepts

Describe the relationship. (There may be more than one.)

BETWEEN...	AND...
anthropocentrism	ecosystem balance
Earth-based spirituality	biocentrism
stewardship ethic	Saint Francis of Assisi
exploitation	wise use
California condor	California condor habitat loss

Did You Know...?

It takes 60,000 pounds of Yew bark (the cutting down of 12,000 trees) to obtain just 2.5 pounds of the cancer-fighting substance taxol.

 Suggested Activities

1. Study the sacred texts of a religion that interests you (if you are a follower of a particular faith, please choose another religion). What does this religion teach about the relationship between humans and nature? Does it consider the natural world to be sacred or inherently valuable as the creation of God? In what ways do its teachings about nature coincide with the teachings of the religion you follow (if applicable)? In what ways do they differ?

2. Become familiar with ecologically appropriate techniques, practices, and devices, and find ways to use them in your own life. For example, start a compost heap in your backyard or neighborhood.

3. Study the life and works of a person who lived in a way that acknowledged the interconnections between humans and other creatures and the nonliving world. Some examples: Saint Francis of Assisi, Aldo Leopold, Rachel Carson, and John Muir.

Chapter 26
Economics and Politics

 Chapter Outline

I. **What Is the Study of Economics?**

   A. Discipline concerned with the production, distribution, and consumption of wealth and with related problems of labor, finance, and taxation

   B. Economics shares same root word (*oikos*, meaning "household") as ecology and literally means "the management of the household."

II. **What Is an Economy?**

   A. Defined as a system of production, distribution, and consumption of economic goods or services

   B. Three Major Types

      1. Pure Traditional Economy

      2. Pure Command Economy

      3. Pure Market Economy

   C. Most, if not all, societies and nations have a mixed economic system.

   D. Goal of most economies: economic growth — increase in the capacity of the economy to produce goods and services

III. **How Is Economic Growth Measured?**

   A. Commonly, via the gross national product (GNP) — total national output of all goods and services valued at market prices in current dollars for a given year

      1. Real GNP — GNP adjusted for any rise in the average price of final goods and services (that is, inflation)

      2. Real GNP per capita — real GNP divided by the total population

      3. GNP is a limited economic measure — tells us nothing about social conditions within a nation.

   B. Alternative indices (to the GNP) factor in environmental and social costs of economic activity.

      1. Net National Product (NNP)

      2. Human Development Index

      3. Index of Sustainable Economic Welfare (ISEW)

IV. How Does Economic Activity Affect the Environment?

A. External Costs

B. Inadequacy of Market Valuation for Ecological Goods and Services

C. Undervaluation of Natural Resources

D. Emphasis on Economic Growth

E. Population Growth

V. What Is Ecological Economics?

A. Transdisciplinary field of study that addresses the relationships between ecosystems and economic systems

B. Goal of ecological economics is sustainability.

C. Sustainability depends on developing a steady-state economy (SSE), one characterized by a constant level of human population and a constant level of artifacts or stock.

D. Sustainable development is defined as improving the quality of human life while living within the carrying capacity of supporting ecosystems.

VI. What Is a Sustainable and Sustaining Earth Society?

VII. How Can Ecological Economics Contribute to Environmental Problem Solving and Management?

A. Environmental Problem Solving

B. Environmental Management

VIII. What Is Politics?

A. Politics, in its broadest sense, encompasses the principles, policies, and programs of government.

B. Environmentalism is a philosophy that transcends party lines.

IX. How Does Politics Affect Environmental Problem Solving and Management?

A. Environmental Problem Solving

B. Environmental Management

X. How Are Political Programs Contributing to Environmentally Sound Management?

A. Convention on International Trade in Endangered Species (CITES)

B. Global Environmental Facility

C. Kyoto Protocol

XI. A Piece of the Antarctic Pie —Exploration Now, Exploitation Later?

A. Antarctic Treaty

Learning Objectives

After learning the material in Chapter 26, you should be able to:

1. Describe the study of economics and explain the three major types of economies.
2. Explain how economic growth occurs and how it is measured.
3. Identify and discuss five factors that render economic activity especially damaging to the environment.
4. Discuss the connection between ecological economics, a steady-state economy, and sustainable development.
5. Explain the relationship between politics and solutions to environmental problems, giving both national and international examples.

Did You Know...?

Environmental degradation can raise a country's GNP, since jobs are created and resources are consumed when (for example) an oil spill is cleaned up, a hazardous waste site is remediated, or a drained wetland is reclaimed and restored.

 ## Key Concepts

Read this summary of Chapter 26 and identify the important concepts discussed in the chapter.

Economics and politics have a significant amount of influence over the resources to which we have access and the measures we use to solve environmental problems and manage resources.

Economics is the discipline concerned with the production, distribution, and consumption of wealth, and with the various related problems of labor, finance, and taxation. In a real sense, it is the study of the management of the household — a state, nation, or the world at large.

All human societies use natural capital (natural resources) and human capital (skill and labor) to produce economic goods, or manufactured capital, such as clothing and appliances. The natural capital (materials and energy) used to produce economic goods is known as throughput.

An economy is a system of production, distribution, and consumption of economic goods or services. In a traditional economy, people grow their own food and make the goods, such as clothing and spears or bows for hunting, that they need to survive. Communities are essentially self-sufficient, and decisions are made locally by individuals and families. Some indigenous groups employ a traditional economy.

A pure command, or centrally planned, economy is one in which the government makes all economic decisions — what goods to manufacture, how much of each good to produce, and how to distribute goods or services among the populace. Few centrally planned economies are still operating today; China and Cuba are notable examples, although the economies of both countries also have aspects of market and traditional economic systems.

In a pure market system, also known as pure capitalism, economic decisions are made by buyers and sellers in the marketplace and are based on the interactions of demand, supply, and price. The market economy of the United States also has aspects of planned and traditional economic systems.

Whatever economic system they adopt, most countries aspire to modernize or develop through economic growth, an increase in the capacity of the economy to produce goods and services. Economic growth occurs through population growth, which may increase the flow of throughput in the economy, increased per capita consumption, or both. It is usually measured in terms of the gross national product (GNP) — the total national output of all goods and services valued at market prices in current dollars for a given year. The GNP includes all personal and government expenditure on goods and services, the value of net exports (exports less imports), and the value of private expenditure on investment. The real GNP is the GNP adjusted for inflation, a rise in the average price of final goods and services. The real GNP per capita, the real GNP divided by the total population, gives some idea of how the average citizen is faring economically. One significant weakness of the GNP is that it fails to account for the environmental degradation and social costs associated with some economic activity. In fact, the GNP provides no information on social and environmental conditions (such as health care, literacy rate, and the degree of environmental degradation) in a nation. As a result, some researchers have attempted to develop better indices of economic activity. These include the net national product, human development index, and index of sustainable economic welfare.

Five factors render economic activity particularly damaging. These are external costs (the harmful social or environmental effects of the production and consumption of an economic good that are not included in the market price of the good), the inadequacy of market valuation for ecological goods and services, the undervaluation of natural resources, an emphasis on economic growth, and population growth.

Ecological economics is a transdisciplinary field of study that addresses the relationships between ecosystems and economic systems in the broadest sense. It integrates many different disciplinary perspectives in order to focus directly on environmental and economic problems. The goal of ecological economics is sustainability, defined as a relationship between dynamic human economic systems and larger dynamic, but normally slower-changing, ecological systems in which human life can continue indefinitely, human individuals can flourish, and human cultures can develop; the effects of human activities remain within limits, so as not to destroy the diversity, complexity, and function of the ecological life support system.

Achieving sustainability requires the development of a steady-state economy (SSE), one characterized by a constant level of human population and a constant level of artifacts, known as stock. The level at which the population and stock are held is sufficient for a good life and sustainable for a long future. The rate of throughput needed to maintain the population and stock is reduced to the lowest feasible level. A SSE is characterized not by growth but by sustainable development, defined as improving the quality of human life while living within the carrying capacity of supporting ecosystems. Sustainable development presupposes sustainable resource use, that is, the use of renewable resources at rates that do not exceed their capacity for renewal. By definition, sustainable use does not apply to nonrenewable resources. At best, the life of nonrenewables can only be extended through recycling, conservation, and substitution.

Sustainable development and a steady-state economy could enable humans to make the transformation from a culture of consumption to a culture of maintenance. In a culture of maintenance, the nurturing instincts of humans could flourish. The resource use model — culture x (resource base/population) = standard of living — enables us to understand disparities in standards of living around the world and to see how we might transform our culture of consumption into a culture of maintenance, and ultimately, a sustainable society. Critical to that transformation is incorporating into our culture an ethic of resource use, an ethic that redefines the human role in nature from conquerors to stewards of the biosphere.

Ecological economics contributes to environmental problem solving and environmental management in numerous ways: improved cost-benefit analyses; true-value pricing that internalizes external costs; the elimination of environmentally harmful subsidies; fees on chemical use, well construction, and other activities in order to protect groundwater; and environmental, or green, taxes.

In its broadest meaning, politics encompasses the principles, policies, and programs of government. Different political systems are based on different principles, and they adopt different policies and programs in order to govern their societies. Even within a particular nation, all citizens do not agree on the best means of government. History has shown that concern for the environment is not the sole domain of any major national political party.

Lack of political will prevents us from alleviating or eliminating many serious environmental problems. The political will to take action on issues usually stems from sufficient public demand. Politics is an important component of international environmental issues as well as national ones.

Many less-developed countries have gone into debt from large-scale development projects such as dams and major roadways. Many are unable to pay even the interest on these loans. However, the loans have not helped improve economic conditions for citizens, and economic woes intensify environmental degradation. Some solutions to the debt crisis are to require that lending institutions act more responsibly with respect to projects that affect the environment; to promote sustainable development; to increase and rechannel development assistance by MDCs to LCDs; to provide grassroots loans to individuals or groups for small, sustainable development projects; and to encourage debt-for-nature swaps. These solutions can help to achieve a sustainable economy, one which maintains its natural resource base, and a sustainable society, a society that works with, not against, natural systems.

Political decisions often affect how resources are managed. By 1995, a relatively small but powerful umbrella group — the self-named "wise use movement" — sought to undo much of the environmental legislation enacted during the past twenty-five years, including the Clean Water Act and Endangered Species Act. "Wise users" claim that federal and state restrictions on what an individual may do with his or her private property are too extreme and in many cases represent an illegal "taking" by the government. They seek an end to these restrictions or compensation for economic losses incurred because they cannot use their land as they see fit. For example, suppose a developer wants to drain a wetland in order to establish residential housing. The wetland provides a multitude of ecosystem services upon which the entire community depends — flood control, the removal of sediments and pollutants from water before it enters a stream or lake, and the replenishment of groundwater supplies. If the wetland is drained, the entire community (including the developer) loses the benefit of those services. To compensate, levies may have to be built to protect against flood waters, the water treatment plant may need to be upgraded to handle the increased load of sediments and pollutants, and so forth. In other words, *all* taxpayers will bear an economic burden because the wetland is drained. Only the developer, however, will directly benefit from the draining of the wetland (through the sale of the homes). Naturally, it is in his or her own self-interest to ignore or minimize the external costs of the drained wetland. By insisting that they be compensated if environmental regulations prohibit them from using their land in certain ways, "wise users" are, in effect, demanding that society pay them not to pollute and/or not to destroy the ecosystem services upon which we all depend.

Antarctica provides one timely example of the impact of politics. The Antarctic Treaty of 1959 was signed to maintain Antarctica for peaceful uses and to promote freedom of scientific research and international cooperation. Twelve nations are parties to the Treaty, which seeks to maintain Antarctic for peaceful purposes and encourage freedom of scientific investigation. Many other nations, particularly LDCs, are arguing for the right to claim the vast natural resources of Antarctica, hoping to increase their economic and political clout. So far, however, research is the only industry on the continent. Some groups, particularly Greenpeace, have called for the establishment of Antarctica as a world park off-limits to war and commerce.

✓ Key Terms

centrally planned economy
culture of consumption
debt-for-nature swap
economic goods
economics
external costs
human capital
index of sustainable economic welfare
manufactured capital
natural capital
politics
pure command economy
real gross national product
steady-state economy
sustainable development
sustainable resource use
throughput
wise use movement

cost-benefit analysis
culture of maintenance
ecological economics
economic growth
economy
gross national product
human development index
internal costs
mixed economic system
net national product
pure capitalism
pure market economy
real gross national product per capita
sustainability
sustainable society
taking
traditional economy

Environmental Success Story

Accommodating rapid growth while preserving environmental quality is a challenge faced by many growing communities; the challenge is particularly pronounced in California's Sierra Nevada region, where rapid population growth is pressuring small towns and villages to spill into the surrounding countryside. Realizing that such sprawl would likely destroy the region's charm, a group of local businesses in 1994 formed the Sierra Business Council to develop a strategy for preserving the environment in which commerce had flourished. Their first publication, the *Sierra Nevada Wealth Index,* identifies 42 indicators of social capital (such as rates of high school completion, volunteerism, and library use) and natural capital (such as agricultural lands, old-growth habitat, and healthy waterways) that make the Sierra Nevada region attractive to business. Through its second publication, *Planning for Prosperity,* and presentations to local leaders, the business council is providing information which helps communities choose alternative forms of development, including mixed-use or transit-centered growth that will help preserve economic and environmental health for generations to come.

Source: Sierra Business Council. "Eighth Annual National Awards for Environmental Sustainability Winners: Urban Sierra Nevada Wealth Index." *Renew America* (22 April 1998). Available: http://solstice.crest.org/sustainable/renew_america/winner98.html. 24 March 1999.

True/False

1. In terms of percentage of GNP, Norway contributes the highest level of non-military assistance to developing countries.   T   F

2. The most serious threat to environmental quality is the global emphasis on economic growth.   T   F

3. Environmental degradation has a positive impact on GNP and GNP per capita.   T   F

4. The Antarctic continent owes its great ice mass to the fact that it receives more than 30 inches of precipitation annually.   T   F

5. All resources can be used on a sustainable basis.   T   F

Fill in the Blank

1. A _____ is characterized by a constant level of human population and a constant level of artifacts.

2. A _____ is based on the concept that one's quality of life should be judged on satisfaction rather than mere possession.

3. Sustainable use does not apply to _____.

4. The only industry in Antarctica is _____.

5. In a(n) _____, a conservation organization "buys" a portion of the debt of a nation at a discount from the bank that made the loan; in exchange, the nation agrees to establish and protect a nature reserve or implement a conservation program.

Multiple Choice

Choose the best answer.

1. The Greek root word for both economics and ecology means

    A. value.

    B. Earth.

    C. household.

    D. pride.

2. The world debt crisis illustrates how environmental degradation and _____ are related

   A. economic chaos

   B. culture

   C. disease

   D. politics

3. _____ is defined as improving the quality of human life while living within the carrying capacity of supporting ecosystems.

   A. Environmental economics

   B. Environmental management

   C. Sustainable development

   D. Sustainable economy

4. According to the "wise use movement," environmental regulations that restrict how an individual may use his or her land amount to a(n) _____ by the federal government.

   A. internal cost

   B. external cost

   C. conspiracy

   D. taking

5. The only kind of resources that have been tapped in Antarctica are

   A. fossil fuels.

   B. minerals.

   C. biological.

   D. fresh water.

Short Answer

1. What is a green tax?

2. Define sustainable resource use. Define sustainable development. How are the two related?

3. What is cost-benefit analysis? What are its strengths and weaknesses?

4. How many countries have claims in Antarctica? As of now, what industries are allowed on the continent?

5. Define traditional economy, pure command economy, and pure market economy; be sure to clearly identify the differences between them.

Thought Questions

Develop a complete answer for each of the following.

1. Discuss the differences between conventional economics and ecological economics.

2. Explain how economics affects environmental management, and give two examples of how economic incentives are being used to encourage environmentally sound management.

3. Describe the impact of large-scale industrial projects on the economic and environmental health of LDCs.

4. Discuss how politics contributes to both environmental problems and their solutions.

5. Detail the limitations of GNP and GNP per capita as indicators of economic welfare, and describe three indices that have been proposed to replace or supplement the GNP.

6. List and describe five ways that economic activity affects the environment. Which do you think is the most significant and why?

7. What is meant by the term "taking" as it is referred to in the Fifth Amendment to the Constitution? What is meant by the term as it is employed by the "wise use movement"? Using recent newspaper and magazine articles, identify five specific cases in which "wise use" proponents claim that health or environmental regulations restrict the way in which a private landowner wants to use his or her land. In each case, what are the external costs associated with the proposed project (or land use)? Are they borne by the community as a whole? What are the benefits to be gained from the proposed land use? Will the entire community benefit? Looking at each case separately, do you think that the restrictions opposed by the "wise use movement" amount to a "taking" by the federal government? Why or why not?

■ Related Concepts

Describe the relationship. (There may be more than one.)

BETWEEN...	AND...
throughout	output
pure command economy	pure market economy
World Bank	world debt crisis
internal costs	external costs
sustainable economy	sustainable society

Did You Know...?

In one $650,000 debt-for-nature swap, Bolivia agreed to protect three million acres of tropical rain forest!

 ## Suggested Activities

1. Join an organization that lobbies to protect the environment.

2. Hold an environmental town meeting. Before a local, state, or national election, invite local candidates to the town meeting to discuss their environmental views.

3. Using either the United States or Canada, compare that nation's economic system with the economic system in place in China. What aspects of a free market system are evident in both economies? What aspects of a centrally planned, or command, system are evident in both economies? You may have to do some reading and research in order to gain a solid understanding of each nation and its economic activities and policies.

4. Find out your legislators' records on environmental issues and write to them to show your support for the environment. You can find out how your member of Congress voted on key environmental legislation by reading *The National Environmental Scorecard,* a voting chart published every two years. The *Scorecard* can be obtained for $15 from the League of Conservation Voters, 2000 L St., NW, Suite 804, Washington, D.C. 20036.

5. Don't forget to vote!

Chapter 27
Law and Dispute Resolution

 Chapter Outline

I. **What Is Environmental Law?**

   A. An organized way of using all of the laws in a nation's legal system to minimize, prevent, punish, or remedy the consequences of actions that damage or threaten the environment

   B. Based on Two Types of Laws: Common and Statutory

      1. Common Law — Written/unwritten principles based on past legal decisions, known as precedents

         a. Nuisance

         b. Trespass

         c. Negligence

      2. Statutory Law

         a. Passed by the state legislature or Congress

         b. Governs how environment and human health are protected and how resources are managed

II. **How Are Environmental Laws Enforced?**

   A. Each law specifies how and by whom it is to be enforced.

   B. Environmental Protection Agency — federal government's primary "environmental watchdog"

   C. Each state has its own enforcement agencies.

   D. Enforcement is not always achieved:

      1. Allocation of authority may be unclear.

      2. Governing agency may disagree with the principle or intent of the law.

      3. Congress/Administration may fail to apportion funds for its enforcement.

      4. Agency may have conflicting priorities.

III. **How Do Environmentally Sound Laws Contribute to Environmental Problem Solving and Management?**

   A. Disputes arise when parties disagree over alternative uses of available resources.

236   Student Learning Guide Chapter 27

B. Environmental Problem Solving — Relevant laws can be guidelines to finding solutions.

C. Environmentally Sound Resource Management — Legal system contributes in subtle ways.

1. Stewardship Ethic — Endangered Species Act

2. Biocentric Worldview — Wilderness Act of 1964

3. Natural System Knowledge — The Food Security Act of 1985

4. Political System Knowledge — Superfund Amendments Reauthorization Act

5. Sociocultural Considerations — National Historic Preservation Act

6. Natural and Social System Research — National Environmental Policy Act (NEPA) and Environmental Impact Statements

7. Economic Analysis — Comprehensive Environmental Response, Compensation, and Liability Act

8. Public Participation — National Forest Management Act

9. Environmental Education — Resource Conservation and Recovery Act

IV. How Is the Field of Environmental Law Changing?

A. Virginia Environmental Endowment

V. What Is Environmental Dispute Resolution?

A. Defined as the process of negotiation and compromise by which disputing parties reach a mutually acceptable solution to a problem

1. Neutral third party, called a mediator, facilitates negotiations.

2. Mediator cannot impose settlements; dispute is settled when parties reach what they consider to be a workable solution.

B. Advantages of Dispute Resolution

1. Gives all parties a better chance to realize objectives

2. Can arrive at solution more quickly

3. Avoids the cost of litigation

C. Disadvantages of Dispute Resolution

1. Inadequate funding for negotiation services

2. Fear of losing power/status by negotiating groups

3. Lack of faith in the outcome of negotiations (outcomes not legally enforceable)

VI. How Does Dispute Resolution Contribute to Environmental Problem Solving and Management?

Learning Objectives

After learning the material in Chapter 27, you should be able to:

1. Define environmental law or legislation.

2. Identify four reasons why environmental laws are not always enforced effectively.

3. Define environmental dispute resolution.

4. Explain how legislation and dispute resolution contribute to environmental problem solving and environmentally sound resource management.

Did You Know...?

Americans throw away enough motor oil each year to fill 120 supertankers.

 Key Concepts

Read this summary of Chapter 27 and identify the important concepts discussed in the chapter.

Environmental law governs the activities of persons, corporations, government agencies, and other public and private groups in order to regulate impacts on the environment and natural resources. Environmental protection resides in common and statutory law.

Common law is a large body of written and unwritten principles and rules based on thousands of past legal decisions dating back to the beginning of the English legal system. It is built on precedent, a legal decision that may serve as an example, reason, or justification for a later decision. Cases involving common law are based on nuisance (a class of wrongs that arise from the unreasonable, unwarrantable, or unlawful use of a person's own property that produces annoyance, inconvenience, or material injury to another), trespass (unwarranted or uninvited entry upon another's property by a person, the person's agent, or an object that he or she caused to be deposited there), or negligence (the failure to exercise the care that "a prudent person" usually takes, resulting in an action or inaction that causes personal harm or property damage). Nuisance, trespass, and negligence are known as torts, causes of action (wrongful acts) for which a civil suit can be brought by an injured plaintiff against a defendant. An injunction is a court order to do or refrain from doing a specified act; an injunction that requires the defendant to stop or restrict a nuisance is called an abatement. Compensation is a monetary award for damages.

Statutory law is the body of law passed by a local legislature or Congress. Congress has enacted several important environmental and resource protection laws. Each environmental law specifies how it is to be enforced and who is to enforce it. The Environmental Protection Agency (EPA) is perhaps the best known of the federal agencies charged with enforcing environmental policies. Each state also has its own agencies.

Environmental laws are not always enforced effectively. Allocation of authority may be unclear. The agency charged with enforcing a law may disagree with the law. The administration or Congress may not provide the funds needed for enforcement. An agency may have a conflict of interest that keeps it from enforcing a law. Finally, the burden of administrative duties may hamper an agency's ability to enforce laws.

Environmentally sound laws are important to environmental problem solving and management. The legal system comes into play at each step of the problem-solving process. Environmental law has advanced significantly since the 1960s, when the first such federal laws were enacted. The earliest laws dealt with such issues as point sources contributing to air and water pollution. Recent laws have focused more on preventing pollution or resource abuse by promoting conservation measures.

Groups such as the Natural Resources Defense Council and creative legal decisions, such as the one that resulted in the creation of the Virginia Environmental Endowment, have helped the field of environmental law to change and grow. Even so, environmental law and litigation cannot provide a satisfactory solution in every case or problem. In such instances, environmental dispute resolution may be helpful. Dispute resolution is the process of negotiation and compromise by which conflicting parties reach a mutually acceptable solution to a problem. A neutral third party called a mediator facilitates negotiations. Dispute resolution offers several important advantages: It hastens resolution, generally costs less, and provides a better chance that all parties will realize their objectives — thus reaching a satisfying solution. Yet dispute resolution does have its disadvantages: Funding for negotiation services is inadequate, some parties fear that they will lose status by appearing to be willing to compromise, and some people lack faith in the outcome of negotiations since they are not legally binding.

Key Terms

common law

dispute resolution

injunction

negligence

precedent

strict liability

trespass

compensation

environmental law

mediator

nuisance

statutory law

tort

Environmental Success Story

Instead of delivering barrels of sludge to corporate headquarters or plugging discharge pipes, Marco Kaltofen has found a more effective way of forcing companies to control their pollution. He lives in the Boston area and operates a chemical analysis laboratory for the National Toxics Campaign. He arms himself with a pH meter, a chemical analysis kit, plastic bags, empty bottles, and a notebook and heads out to gather information. When he is ready with his results, he publicizes his findings by calling press conferences. As a result of one of his successful investigations, American Cyanide pleaded guilty to 37 criminal counts of polluting the Rahway River and was fined $900,000. Because of the slow nature of bureaucracy, Kaltofen believes that citizens are more effective than government agencies at enforcing environmental laws.

True/False

1. A person can be charged with trespass only if he or she enters another person's property. T F

2. If harm results from a product, the maker of that product is not liable for the harm done if it is determined to have been unforeseeable. T F

3. Statutory laws are derived from precedent. T F

4. Dispute resolution relies upon negotiation and compromise. T F

5. A legal decision that serves as an example, reason, or justification for a later decision is known as a prior restraint. T F

Fill in the Blank

1. Two methods for settling environmental disputes are _____ and _____.

2. In 1970, President Nixon created the _____ to enforce environmental laws.

3. The most common cause of action in environmental law is _____.

4. In dispute resolution, the disputing parties are assisted by a neutral third party, called a(n) _____.

5. A large body of written and unwritten principles and rules based on thousands of past legal decisions dating back to the beginning of the English legal system is _____.

Multiple Choice

Choose the best answer.

1. A monetary award for damages is called a(n)

    A. fee.

    B. litigation.

    C. abatement.

    D. compensation.

2. The failure to exercise care that "a prudent person" usually takes, resulting in an action or inaction that causes personal harm or property damage, is called

    A. nuisance.

    B. trespass.

    C. negligence.

    D. liability.

3. A court order to do or refrain from doing a specified action is a(n)

    A. legal writ.

    B. abatement.

    C. precedent.

    D. injunction.

4. The process of negotiation and compromise by which opposing factions attempt to resolve a problem is called

    A. statutory law.

    B. teamwork.

    C. common law.

    D. dispute resolution.

Short Answer

1. What is an abatement and what is it an example of?
2. List the three causes of action upon which cases involving common law are based.
3. Define nuisance and give an example.
4. Define negligence and give an example.
5. What is coordinated resource management planning?

Thought Questions

Develop a complete answer for each of the following.

1. Why are environmental laws not always effectively enforced?
2. Discuss how the legal system contributes to environmentally sound resource management.
3. What role does law play in environmental problem solving?
4. Reread the section on the Virginia Environmental Endowment and let it inspire you. Now, brainstorm some positive, creative, and constructive ways that polluters might be made to compensate society for the pollution they have caused.
5. Define dispute resolution and explain why many people are beginning to use it as an alternative to litigation. What are its advantages and disadvantages?

▣▬■ Related Concepts

Describe the relationship. (There may be more than one.)

BETWEEN...	AND...
common law	statutory law
biocentric worldview	Wilderness Act of 1964
stewardship ethic	Endangered Species Act
negligence	liability
dispute resolution	coordinated resource management planning

Did You Know...?

The Endangered Species Act is considered by many people to be the most important and progressive piece of environmental legislation ever written.

 ## Suggested Activities

1. Find out who enforces environmental laws in your county or state and what means they use to ensure compliance.
2. Join a public interest organization that litigates on environmental issues.
3. Find out more about the programs developed by the Virginia Environmental Endowment and look for opportunities to apply them to problems in your area.

4. Try this role playing exercise: Form a coordinated resource management planning team with other members of your class. Adopting the roles of different types of resource managers and users, work together to develop a management plan for the resource of your choice.

Chapter 28
Environmental Education

 Chapter Outline

I. **What Is Environmental Education?**

   A. Definition

   1. Goals as stated in the first issue of *Environmental Education*:

      a. Aimed at producing citizenry that is knowledgeable concerning the environment

      b. Should make people aware of environmental problems and how to help solve them

      c. Should motivate people to work toward their solutions

   2. Goals outlined in 1977 by the World Intergovernmental Conference:

      a. To develop citizenry that is aware of and concerned about the environment

      b. To foster knowledge, motivation, attitudes, skills, and commitment to work toward solutions to problems

      c. To encourage people to apply same level of commitment to preventing new problems

   B. Majority of EE programs foster environmental awareness rather than problem solving.

   C. Broad goals for curriculum development in environmental education:

   1. Understanding of how nature works

   2. Awareness of the interdependence of all living and non-living components of the biosphere

   3. Understanding of how environmental problems are caused

   4. An interdisciplinary approach to both education and environmental problem solving

   5. Belief that education should lead people to become involved in environmental issues should they choose to do so

II. **What Kinds of Environmental Education Programs Are Available?**

   A. Institute For Earth Education (IEE)

B. Roots and Shoots

C. Project WILD

D. School For Field Studies (SFS)

E. Sierra Institute

F. Au Sable Institute of Environmental Studies

III. How Is Environmental Education Taught Formally?

A. Teaching EE as a separate course

B. Integrating EE into all courses

IV. How Is Environmental Education Taught Informally?

A. Examples of organizations/forums (such as zoos and nature centers)

B. Informal educational programs are also offered by national and international environmental organizations.

C. EE is more than a passing movement, but more effort is needed to weave it throughout social fabric.

V. How Does Environmental Education Contribute to Environmental Problem Solving and Management?

A. Considers the environment as a whole

B. Increases awareness of global problems

C. Emphasizes the role of values, morality, and ethics in shaping attitudes and actions affecting the environment

VI. Environmental Education: Where Do We Go From Here?

Learning Objectives

After learning the material in Chapter 28, you should be able to:

1. Discuss what is included in environmental education.

2. Explain the difference between formal and informal environmental education.

3. Describe two approaches to integrating environmental science into school curricula.

4. Explain how environmental education affects environmental problem solving and management.

Did You Know...?

Kids For a Clean Environment (FACE) was founded in 1989 by a nine-year-old girl as an after-school club at her school. Today, it has over 300,000 members worldwide.

 Key Concepts

Read this summary of Chapter 28 and identify the important concepts discussed in the chapter.

Environmental education seeks to awaken and explore a person's values. It is aimed at producing a citizenry knowledgeable about the environment and its associated problems, aware of how to help solve those problems, and motivated to work toward their solutions.

Environmental education differs from ecology and science education in that it contains a strong social component. Five elements are especially important to environmental education programs: an understanding of, and appreciation for, how nature works; an awareness of the interdependence of all life and of the living and non-living components of the biosphere; an understanding of how environmental problems are caused (through the interaction of natural and cultural systems); an interdisciplinary approach; and action (so that people can become involved in environmental issues should they choose to do so).

There are a wide variety of environmental education programs available at the local, national, and international levels. They are offered through schools, nature centers, non-profit organizations, and other entities. In a formal setting, there are at least two approaches to environmental education. Some teachers advocate adding a separate environmental course or courses to the curriculum. Others advocate making environmental issues an integral part of all courses in the curriculum. Doing so allows students to see the connection between environmental issues and other topics.

Many groups outside of the formal school system are important targets for environmental education because their action or inaction can influence local environmental quality. Informal programs encompass a broad array of educational approaches at diverse sites such as zoos, aquaria, nature centers, parks, and wildlife refuges. As with formal educational programs, informal programs are increasingly developed around problem-oriented objectives.

Environmental education contributes to environmental problem solving and management in many ways: by highlighting the interaction between natural and cultural systems; by developing critical thinking and problem solving skills; by increasing awareness of global problems and issues while facilitating local action; and by emphasizing the role that values, morality, and ethics play in shaping one's attitudes and actions toward the environment.

Environmental education that connects children and adults to nature in meaningful, nonthreatening ways can serve as an antidote to "ecophobia," the fear that some people feel when confronted with environmental problems such as climate change, stratospheric ozone depletion, and deforestation. The basis for meaningful environmental education is nature study, which satisfies students' curiosity about the world around them and encourages them to get outside where they can learn and enjoy simultaneously.

Environmental Success Story

While the American public laments about the low math and science test scores of its high school students, Washington, D.C.'s Environmentors Project is proving that high school students can and do take the sciences seriously. Through the project, low-income public high school students conduct environmental research and community service projects while being mentored by environmental professionals and scientists. The pairs work together once each week for four months in the mentors' laboratories, offices and "the field," where students design and create projects that identify real environmental problems in their communities and present potential solutions. The project also sponsors monthly educational trips and seminars which raise students' awareness of the "bigger picture" and help them learn about environmental and career options. At the end of each session, students present their projects at an environmental science fair and compete for college scholarship awards. The Environmentors Project also works to provide other cities with the information necessary to create similar programs in their communities.

Source: The Environmentors Project. "Seventh Annual National Awards for Environmental Sustainability Winners: The Environmentors Project." *Renew America* (22 April 22 1998). Available: http://solstice.crest.org/environment/renew_america/winners.html. 24 March 1999.

True/False

1. All states require environmental education to be taught in secondary schools.   T   F

2. Environmental education cannot be successfully integrated into regular classroom courses.   T   F

3. One disadvantage of having separate environmental education courses is that students may not be able to see how the topics relate to the rest of their lives.   T   F

4. Many national and international conservation organizations offer informal education programs.   T   F

5. Environmental education, if properly taught, can be limited to school children and those in college; it is not needed by the general populace.   T   F

Fill in the Blank

1. _____ seeks to awaken and explore a person's values; it is aimed at producing a citizenry knowledgeable about the environment and its associated problems, aware of how to help solve those problems, and motivated to work toward their solution.

2. The _____ is an international, nonprofit, volunteer organization whose motto, "Learning to Live Lightly," reflects its desire to help people examine and change their lifestyle.

3. The Rachel Carson Project in the Corvallis, Oregon, school system incorporates environmental topics into these four standard high school courses: _____, _____, _____, and _____.

4. Two states mandate environmental education in schools; they are _____ and _____.

Multiple Choice

Choose the best answer.

1. In what decade did the concept of values become part of the definition of environmental education?

    A. 1960s

    B. 1970s

    C. 1980s

    D. 1990s

2. One of the goals established by Project WILD includes:

    A. awareness.

    B. skills.

    C. knowledge.

    D. All of the above are true.

3. Which of the following is a program for college students?

    A. Sierra Institute

    B. School for Field Studies

    C. Sunship Earth

    D. Earthkeepers

4. "Ecophobia" is a term that refers to:

   A. a fear of housekeeping.

   B. an irrational fear of environmentalists.

   C. fear of some people when confronted with environmental problems.

   D. None of the above is true.

Short Answer

1. Briefly describe two approaches to adding environmental education to a school curriculum.

2. Describe the target groups for informal environmental education and briefly explain why it is important that these groups become environmentally aware.

3. What three project areas does the Roots and Shoots program deal with?

4. Name at least five forums for informal environmental education.

Thought Questions

Develop a complete answer for each of the following.

1. Describe the goals of environmental education and give examples of ways environmental educators are trying to achieve these goals.

2. Who is contributing to informal environmental education efforts? Describe the kinds of things these groups are doing to make people more environmentally aware.

3. Why are values an important part of environmental education?

4. Discuss how environmental education could be integrated into the following courses: Literature, History, Art, Math, Geography, Social Studies, Physics, and Chemistry.

5. Differentiate between formal and informal environmental education. What are the advantages and disadvantages of each? Which do you think is most effective and why?

▢▨■ Related Concepts

Describe the relationship. (There may be more than one.)

BETWEEN...	AND...
formal environmental education	informal environmental education
global ethic	environmental education
discrete environmental course	environmental education integrated into other courses
environmental education	a sense of wonder

Did You Know...?

In just 20 years, the IEE has created six module Earth education programs, developed 250 educational activities, and hosted 1,500 workshops and training sessions.

 ## Suggested Activities

1. Urge the PTA of a local school to set up an environment committee and endorse an environmental curriculum.

2. Conduct a campus-wide environmental audit to identify opportunities to minimize waste generation, reduce water and air pollution, and conserve energy and water.

3. Urge your student government to pass resolutions supporting environmentalism.

4. Create a coalition of campus groups to promote environmental issues.

5. Organize an environmental teach-in on Earth Day (April 22).

6. Try the exercise in Box 28-1: *Developing A Sense of Place*, page 594, in the textbook. How much do you know about the region in which you live? How can you go about increasing your sense of place?

Special Supplement:
Guidelines for Research

Introduction

This section of the *Student Learning Guide* was written to help students who have been given the assignment of a research paper, term paper, or other research project in environmental problem solving and environmentally sound management. Although most students probably had some experience with this in high school, many students get to college without ever having performed a research project and are intimidated by the prospect of one. If you are in this situation now, you may have a lot of questions: How do I choose a subject? How do I use the library? Where do I look for information? How much information is enough? What should I discuss in my research paper?

Don't panic. This section gives answers to these and other questions in the form of guidelines. If you follow these guidelines, you will be able to conduct research successfully.

This section gives you a thorough, step-by-step process for researching, but it is not exhaustive. You will undoubtedly discover other sources of information specific to your topic. Don't hesitate to use them. For example, as you read an article on radioactive waste disposal in *New Scientist,* you find the name of a prominent scientist who is an expert on the subject. Find out where she can be reached, and contact her to request an interview.

As the preceding example suggests, the research process does not necessarily begin and end in the library. Some of your most valuable sources of information will be your own experiences and the experiences of other people — especially experts on your subject.

Most of the reference materials mentioned in these guidelines are fairly common and easy to find at college and university libraries. In addition, almost every library has many resources not mentioned here. Work closely with your library staff to locate useful materials.

Choose a Subject

The first task in your research strategy is choosing a subject to research. The subject you choose will determine how and where you will look for the information, and how much information you will find. Your subject should be:

- **Of Interest to You**. Would you like to learn about the economic effects of a fossil fuel shortage? Are you interested in public education on the loss of tropical rain forests? Do you want to investigate the health effects of toxic waste? Your enthusiasm will take you a long way toward a successful project.

- **Worthy of Research**. If you choose a subject that is trivial, you won't be able to find enough information on it.

- **Specific**. Too broad a topic will result in information overload and badly organized research. For example, acid precipitation is a global problem, so "acid rain" is probably too broad a subject. Some initial investigation into the topic, though, will reveal that acid rain is a particular problem in the Adirondacks region of the United States. Acid rain in the Adirondacks is a better subject.

- **Not Too Specific**. If your subject is obscure or narrow, your library may not have any information on it. For example, the effects of a particular chemical on a particular species is probably too specific, but the effects of toxic chemicals on a particular ecosystem is probably not.

How do you know if you've chosen a good subject? Your instructor and your librarian are good people to discuss it with. They will have a good idea of the information that is available and whether or not your subject is too broad or too narrow.

After you have chosen a subject, head for the library and read general discussions on the resource. A good place to start is with the science dictionaries and textbooks. Science encyclopedias, such as the *McGraw-Hill Encyclopedia of Environmental Science* and the *McGraw-Hill Encyclopedia of Science and Technology,* can be found in the reference section. Read the information on that topic in your textbook. Reference materials (encyclopedias) and general materials (textbooks) will give you a good overview of the issues related to your subject and will help you narrow it down to a more workable size, if necessary.

Don't overlook two very important sources of help at this stage: your instructor and your librarian. Both are well-qualified to help you in your search.

Use these and other books (including textbooks) to gain a basic understanding of your subject:

- *Grzimek's Encyclopedia of Ecology*. Bernhard Grzimek, Joachim Illies, Wolfgang Klausewitz, editors.

    A collection of essays that provides a good introduction to various environmental topics.

- *McGraw-Hill Encyclopedia of Environmental Science*. New York: McGraw-Hill Book Co.

    A good collection of introductory essays.

Develop an Outline

After you've chosen a subject, develop an outline. We suggest using the same approach as that used in the "Environmental Science In Action" web pages that accompany *Biosphere 2000: Protecting Our Global Environment* (Available: http://miavx1.acs.muohio.edu/~kaufmadg/).

I. **Describing the Resource**. Begin with a comprehensive description of the resource. Describe all physical, biological and social characteristics, such as resource availability, wildlife, and cultural beliefs and values.

II. **Looking Back**. Provide a historical account of previous management practices. Include an analysis of successful and unsuccessful practices. Your research for this section will call for investigation into the technologies and government policies associated with the resource.

III. **Looking Ahead**. Preview what lies ahead for the resource or natural system you have chosen. Develop a plan for future management. This is the heart of environmentally sound management. Your personal discovery is worthwhile because as you develop a management plan, you will begin to understand what needs to be done, on a personal as well as societal level, to manage resources wisely.

The outline doesn't have to be detailed at this point — just enough to help you organize your research. You probably will add a second level of detail after reading a couple of encyclopedia articles or textbooks. You'll add additional levels of detail as you get more information.

Begin Your Research

Once you have chosen a subject, narrowed it down to a manageable size and developed an outline, you are ready to begin your research. The following information describes how to locate the primary types of research materials found in libraries: books and periodicals.

Library of Congress Subject Headings

The Library of Congress publishes a list of all the **subject headings** used in libraries. This list is the key to using the card catalog, computerized retrieval systems, and many other indexing tools because it tells you the correct subject headings to use. Ask your librarian to help you use it. The list will save you a lot of time and frustration if you use it before you use the card catalog!

For example, if you choose to research endangered species, you would look under the following headings:

EXTINCT ANIMALS

RARE ANIMALS

RARE BIRDS

WILDLIFE CONSERVATION

PLANT CONSERVATION

RARE PLANTS

Card Catalog

The card catalog is a valuable tool for locating books and reports. It is the index to the books that the library owns, and it provides information on where these books can be found on the shelves. You can locate this information by using the three types of cards in the catalog — author, title and subject heading cards. In your early research, you will probably not know any specific titles or authors and will need to look up books by the subject heading. The subject headings are the phrases you looked up in the Library of Congress Subject Headings.

Periodical Indexes

Books provide useful information, but they take a while to write and publish. If you want to find the most recent information (and you do), then you need to look for journal articles. Many journals are published weekly or monthly, so their information is very up-to-date. A **periodical index** tells you what articles have been published and what journals have published them. Each index is arranged a little differently, but all of them give you the information you need to locate the article you want: author, article title, journal title, volume number, page numbers, and date of publication. This information is called the **citation.**

Abstracted indexes can save you a lot of time and effort. They provide the same essential information as the other indexes, but they also provide summaries (abstracts) of the articles indexed. Read these summaries and decide whether or not reading the entire article would aid your research. The index section provides the citation information you'll need to locate the original article.

Periodical indexes frequently have different index sections, such as subject and author indexes. You will probably use subject indexes most often. Which section you use depends on your research needs. For example, if you are looking for information from only one region, see if the index has a geographic index. The same is true if you are searching for a specific author — look for an author index. When you have located a journal article that you want to read, check your library's **serials holdings list** — the list of all the periodicals that the library subscribes to — and find out if that journal is available. Ask your librarian for the holdings list.

Several of the more useful indexes are listed below.

- *Applied Science and Technology Index.* Bronx: The H.W. Wilson Co.

    This index covers a variety of science and mathematics periodicals. It is most useful for research on energy resources.

- *Environment Index* and *Environment Abstracts*. New York: EIC/Intelligence.

    These two books have indexes and abstracts (summaries) of key environmental literature each month. You can use them to look up subjects, authors, and geographic areas. Use the *Index* to locate the citations that deal with your subject. Then look in the *Abstracts* for summaries of those sources. If you would like to locate the original article, the books provide the information you need. The *Index* also has a Review Section which contains essays on major environmental events and problems, a directory of contacts in federal and state agencies, regional commissions and non-governmental associations, and a list of recent environmental books and films.

- *General Science Index.* Bronx: The H.W. Wilson Co.

    As the title implies, this is a general index which covers a wide range of topics, including atmospheric science, biology, environment and conservation, food and nutrition, oceanography, and zoology.

More Periodical Indexes

Some periodical indexes aren't solely for environmental information, but they may still be helpful. A few of these indexes are listed below. The periodical index you choose will depend on the aspect of management you are investigating. For example, if you want to know more about the attitudes of the current administration in Washington, D.C., toward the environment or the economic feasibility of alternative energy sources, you might look in the *Business Periodicals Index*.

Other indexes you can consult are:

- *Education Index*
- *Public Affairs Information Service* (PAIS)

Bibliographies

Many books and articles have **bibliographies** which will give you additional places to look for information. Always check for a bibliography in any source you use. Then follow up on any books or articles you think might be helpful in your research.

Government Documents

The United States Government is the world's largest publisher. The federal government sponsors a lot of research and publishes the results. These public documents are an invaluable source of information. Government agencies relevant to the environment include the Environmental Protection Agency, the United States Geological Survey, the Department of Energy, the Department of Agriculture, and the Department of the Interior. Publications are listed in the *Monthly Catalog of U.S. Government Publications*, the main index to all U.S. executive, judicial, and Congressional publications.

All libraries, even the smallest public library, have access to government documents. Some libraries receive government documents regularly — these are the **depositories.** Check with your librarian to find out where the nearest U.S. government document depository is. The regional depository is responsible for making the documents available to other libraries. Your library will contact the depository on your behalf.

NOTE: Government documents are arranged under the Superintendent of Documents (SuDoc) classification system. Reports are classified by the agency which issued them. All Department of Agriculture publications are under the letter A; publications of the Environmental Protection Agency are together under EP. Ask the librarian for assistance.

Other guides to government publications are listed below:

- *American Statistics Index and Abstracts* (ASI). Washington, D.C.: Congressional Information Service, Inc.

    If you need statistical information for the United States, ASI can help you find it. You can use these books to locate information from the many statistical sources published within the federal government. (International statistics are published in several publications of the United Nations.)

- *Monthly Catalog of U.S. Government Publications.* Washington, D.C.: U.S. Government Printing Office.

    This index covers almost all government publications, including House and Senate hearings, bills and laws, and research reports. The Environmental Protection Agency publishes many reports on environmental research; the Department of Energy and the Department of Agriculture also publish relevant materials. Other useful materials are published by the National Park Service and U. S. Forestry Service.

- *Congressional Information Service* (CIS Index). Washington, D.C.: Congressional Information Service, Inc.

    The CIS Index provides information (including abstracts) on Congressional reports. You can use these to find information on environmental legislation.

- *National Technical Information Service (NTIS Reports* indexes)

    NTIS reports are government-funded research reports. These reports come from universities, corporations, and research foundations and cover many subject areas. Use the *Government Reports Announcements and Index* and the *Government Reports Annual Index* to identify relevant and useful reports. If your library does not own the report, you may be able to order it. Ask the documents librarian for information.

United Nations Publications

The United Nations publishes many documents which provide information on foreign countries. The UNDOC indexes all official and non-official publications. Some useful indexes are:

- *Directory of International Statistics*
- *World Population Trends*

- *World Population Prospects*
- *Handbook of Industrial Statistics*
- *Energy Statistics Yearbook*

Other Sources

- *Dictionary of the Environment.* New York: New York University Press.

  This helpful dictionary defines terms pertaining to environmental science.

- *Environmental Glossary.* Rockville, MD: Government Institutes, Inc.

  This glossary defines the terms which are used in federal statutes and also provides an explanation of basic regulatory concepts used by the Environmental Protection Agency.

Newspapers

Don't overlook newspapers as a source for recent and regional information. Ask the librarian to help you with the newspaper files. Find articles on local issues and events which concern your subject. Was there a chemical spill in your area? Find the news reports on it. Did your hometown recently institute a pollution clean-up plan? Look for the local press coverage.

For issues on a national or international level, the *New York Times Index* is a good place to start.

Take Notes

As you read about your subject, take notes. Organize your notes according to your outline. A good way to manage this is to keep a notebook with tabs for each of your major headings, and put your notes behind the appropriate tab.

HELPFUL HINT: Don't write the author's exact words in your notebook in your own handwriting. If you do, when the time comes to write about it, you may not remember whether you paraphrased or not. If you think you may want to quote a source, photocopy that page and put it into your notebook. Otherwise, always paraphrase the information you jot down in your notes. It will make it a lot easier when you start writing, and you won't inadvertently plagiarize.

Cite Your Sources

As you take notes, be sure to keep careful track of the reference materials you use. You will need to give credit to all of the sources you use in your research, according to the citation rules that your instructor prefers. If you wait until you are ready to type your paper to begin recording citations, chances are you will forget some of your sources. You will have to make several return trips to the library to look up missing information.

To avoid having to do your research over again, keep information on all your sources in your notebook. For books, record the author, title, date of publication, publisher, city, and page numbers. For journal articles, record the author, article title, journal title, volume number, issue number, page numbers, and date. You don't have to worry about the format of this information now, but you'll be glad that you have it when the time comes to type it later.

If you plan to use direct quotes, be sure to record them accurately. A good way to ensure that you've gotten the quote right is to photocopy the section from the original text. The more you recopy text by hand (or by word processor), the more likely you are to incorporate errors into the quotation.

IMPORTANT: You must give citations for any information that you use in your paper. You are probably aware that if you use direct quotes, you must provide references to the sources. You must also provide citations for paraphrased material. If you don't give proper credit, you will be guilty of plagiarism.

Sources Outside the Library

There is a wealth of information which cannot be found on a bookshelf. By venturing outside the library, you not only can get information relating to your subject, but you also will get an interesting view of various types of organizations — public and private. Many groups will supply you with materials for free or for a nominal fee.

Four sources are especially important:

- Individuals with special interests in environmental science and environmentally sound management
- Environmental organizations
- Governmental organizations
- Private environmental research and management firms

Contacting Individuals with Special Interests

There are two ways that individuals with special interests may aid you in your research. They may act as sources for information (either through an interview or by providing you with written information) or they may act as reviewers, checking the information that you have compiled.

Many environmental scientists and resource managers who are concerned about the environment are very willing to help students learn about their areas of expertise. But remember that their time is valuable. In other words, be prepared and as well-informed as possible before you approach them. Through a letter or a brief phone call, you can request information or an interview. Tell them that you are a student who is interested in environmental management and explain why you would like information.

If the person is willing to be interviewed, arrange a time for an interview, either over the phone or in person. Prepare a list of questions to guide the interview (but don't limit yourself to these). Take notes during the interview and, if possible, tape the session to transcribe later. If an interview is not possible, request written information. The person may be willing to send you information that is not available at your library.

Contacting Environmental and Governmental Organizations

Environmental and governmental organizations are invaluable sources of information. Write or call for printed materials on your subject. The Sierra Club, The Nature Conservancy, World Wildlife Fund, and The Conservation Foundation are just some of the environmental organizations which may provide you with useful information. The Environmental Protection Agency, the Department of Agriculture, and the Department of Energy are only a few of the many helpful government organizations. The United Nations can also supply information.

At the state level, you can contact the state Environmental Protection Agency and natural resources and health departments.

Contacting Private Research and Resource Management Firms

Many private firms conduct research related to the environment and can provide you with information on their work. Locate firms which are involved with your resource and contact them. If, for example, you are researching the handling of radioactive materials, you might contact a company such as Westinghouse, which performs research in this area. Request information which could help you form your management plan. (Be aware, however, that some information is proprietary, and firms will not release it.)

Write Your Paper

With your research done, you're ready to write. Follow the outline you used for your research. Find out from your instructor if he or she has a particular style preference for such things as footnotes and bibliographic citations. If not, we suggest you purchase an inexpensive style guide for research papers. Several should be available at your campus bookstore. Most of these books also provide writing tips. When writing your first draft, include all the information that comes to mind. If your paper is too long, you can go back later and edit. Revise your work until you are satisfied.

The hard part is over. At this point, you just need to make sure that the paper you hand in reflects your hard work. If you use word processing software to type your paper, use a spell checker to correct typographical errors and misspellings. To make doubly sure you catch errors, proofread a printed copy. Better yet, have a friend proofread it.

Finally, turn it in...on time.

Answer Section

Chapter 1

True/False

1. F
2. F
3. F
4. T
5. F

Fill in the Blank

1. 20 percent
2. pollutant
3. anthropocentric
4. stewards (responsible for caring for the Earth)
5. industrialization and urbanization

Multiple Choice

1. B
2. D
3. D
4. C
5. A

Short Answer

1. Spaceship Earth is an analogy used by environmentalists in the 1960s and 1970s that described the Earth as a closed system with finite resources and limited ability to recycle pollution.

2. A biocentric worldview sees humans and human culture as part of nature; an anthropocentric worldview sees humans and human culture as separate from nature.

3. Organisms live at the expense of their environment, have cellular structure, reproduce, respond to stimuli, show growth, evolve and adapt, and exhibit movement.

4. Renewable resources are those that can be replaced by the environment if they are not used up faster than they can be restored. Nonrenewable resources are those that are finite in supply or are replaced so slowly that they may as well be finite.

5. Net primary productivity is the total amount of solar energy fixed biologically through photosynthesis minus the amount of energy that plants use for their own needs.

Chapter 2

True/False
1. F
2. F
3. F
4. F
5. F

Fill in the Blank
1. ecology
2. scientific inquiry
3. ecosystem
4. litigation
5. stewardship ethic

Multiple Choice
1. C
2. C
3. B
4. C
5. D

Short Answer
1. Scientific inquiry is the process of observation, hypothesis development, and experimentation.
2. Scientists ensure that their research is free of bias or coercion by using controlled experiments and reporting methods and results to other scientists (who may repeat the experiment).

3. An ecosystem is a self-sustaining community of organisms interacting with one another and with the physical environment within a given geographic area.

4. Ecology is important for understanding the structure and function of ecosystems. Theology, education, and art are important for the shaping of worldviews. Theology helps us to understand human motivations and to answer questions about our origins. Education shapes opinions and perceptions and is a means of sharing information. Art can encourage or discourage stewardship by expressing the value placed on the natural world and the relationship between natural and cultural systems.

5. The five steps of the environmental problem-solving model are as follows:

   1) Identify and diagnose the problem.

   2) Set goals and objectives.

   3) Design and conduct a study.

   4) Implement, monitor, and re-evaluate the chosen solution.

Chapter 3

True/False

1. F
2. F
3. F
4. T
5. T

Fill in the Blank

1. habitat
2. abiota and biota
3. compounds
4. macronutrients
5. primary consumers (herbivores)
6. detritivores (detritus feeders)

Multiple Choice

1. B
2. D

3. D
4. A
5. B

Short Answer

1. Ecology is studied at the individual, species, population, community, ecosystem, biome, and biosphere levels.

2. Entropy is the flow of energy from a high-quality, concentrated and organized form to a low-quality, randomly dispersed and disorderly form.

3. Elements are substances that cannot be changed to simpler substances by chemical means.

4. The six macronutrients are carbon, oxygen, hydrogen, nitrogen, phosphorus, and sulfur.

5. Eutrophication is the natural aging of a body of water caused by nutrient enrichment.

Chapter 4

True/False

1. T
2. F
3. T
4. F
5. F

Fill In the Blank

1. gross primary productivity
2. respiration
3. nitrogen fixation
4. carbon, nitrogen, sulfur, and phosphorus
5. global warming

Multiple Choice

1. A
2. D

3. C
4. C

Short Answer

1. Respiration is the release of energy from fuel molecules.
2. The pyramid of energy is a model that depicts the production, use, and transfer of energy from one trophic level to another.
3. The pyramid of numbers is a model that depicts the relative abundance of organisms at each trophic level; carnivores at the top are relatively few in number.
4. The pyramid of biomass is a model that depicts the relative amounts of biomass at each trophic level. There is less biomass at higher trophic levels. (The organisms are larger in size, but there are fewer of them.)
5. Trees and other plants use carbon dioxide in the process of photosynthesis. Oceans dissolve carbon dioxide, and there it may combine with calcium or magnesium to eventually form limestone or dolomite.
6. The gross primary productivity (GPP) is the total amount of energy fixed by autotrophs over a given period of time. The net primary productivity (NPP) is the GPP minus the amount of energy that autotrophs use to meet their own energy needs. The NPP is the amount of energy available to other organisms (heterotrophs) in the community.

Chapter 5

True/False

1. T
2. T
3. F
4. F
5. T

Fill in the Blank

1. ecosystem succession/ecosystem development
2. inertia
3. endoparasite
4. lag phase
5. carrying capacity

Multiple Choice

1. A
2. C
3. D
4. A
5. D

Short Answer

1. Stress is any change in the environment of an ecosystem. It is a disturbance that alters ecosystems. Examples include climatic changes, natural disasters, and human activities.

2. A climax community is an association of organisms best adapted to the physical conditions of a geographic area; the community is usually dominated by a few abundant plant species, such as a beech-maple forest.

3. Primary succession occurs where no organisms previously existed (such as on bare rock). Secondary succession occurs where an existing ecosystem has been disturbed.

4. Pioneer organisms begin the process of soil formation in primary succession. Their organic detritus mixes with bits of weathered rock. The primitive soil they create enables other organisms to establish themselves.

5. The competitive exclusion principle is the phenomenon that operates when interspecific competition leads to the exclusion of one of the competing species.

Chapter 6

True/False

1. T
2. T
3. F
4. F
5. T

Fill in the Blank

1. damage, disruption, destruction, desertification, deforestation
2. disruption

Student Learning Guide Answer Section   **265**

3. destruction
4. point
5. cross-media pollutant

Multiple Choice

1. B
2. D
3. A
4. B
5. D

Short Answer

1. The four factors that determine how a pollutant affects the environment are (1) the effect of the pollutant, (2) how it enters the environment, (3) the quantity discharged to the environment, and (4) its persistence.

2. Synergism occurs when substances interact to create an effect greater than the sum of their individual effects. For example, photochemical smog, which forms when hydrocarbons and nitrous oxides mix and are exposed to sunlight, contains compounds which are more harmful than the original components combined.

3. Rangelands and croplands (especially semi-arid and arid) are most vulnerable to desertification because these ecosystems contain vegetation well-suited to foraging by animals and because intense agricultural activity magnifies the effects of stresses brought on by lack of moisture.

4. The major causes of deforestation are the conversion of land to agricultural use or pastureland; the demand for fuel, timber, and paper products; and the construction of roadways.

5. Preventative desertification measures must be site-specific and match local land, climate, economic, and cultural conditions.

Chapter 7

True/False

1. F
2. T
3. T
4. F

5. T

Fill in the Blank

1. disturbance ecology
2. landscape ecology
3. indicator species
4. agroecology
5. conservation, preservation

Multiple Choice

1. B
2. C
3. B
4. D
5. A

Short Answer

1. The five subdisciplines of applied ecology are disturbance ecology, restoration ecology, landscape ecology, agroecology, and ecotoxicology.

2. Applied ecology involves using scientific knowledge to predict the ecological impacts of human activity and *manage* natural systems. Therefore, it must incorporate societal values. Ecology, on the other hand, strictly enhances knowledge about structure and function of communities.

3. The three major goals of restoration ecology are (1) to repair biotic communities after a disturbance, (2) to maintain present diversity of species and ecosystems, (3) to increase knowledge of biotic communities and restoration techniques.

4. Conservation biology is the scientific discipline dedicated to protecting, maintaining, and restoring the Earth's biological diversity.

5. Agroecosystems are less complex and diverse than natural ecosystems; they also require inputs of water, energy, and nutrients.

Chapter 8

True/False

1. T
2. F

3. T
4. F
5. F

Fill in the Blank

1. crude birth rate
2. replacement fertility
3. population momentum
4. gross national product (GNP)
5. infant mortality rate

Multiple Choice

1. C
2. C
3. D
4. A
5. C

Short Answer

1. The actual population increase in absolute numbers for a particular country or region is determined by the following equation: actual population increase = (births + immigration) - (deaths + emigration).

2. Zero population growth (ZPG) occurs when the number of deaths equals the number of births, and immigration equals emigration (that is, no growth occurs in absolute numbers).

3. The doubling time is the number of years it will take for a population to double in size. It is calculated by the rule of 70: 70/growth rate = number of years to double (assuming growth rate remains constant). The doubling time is a useful measure because it allows us to visualize what future environmental and social conditions would be like if present growth rates were to continue.

4. The three most important measures of fertility are the general fertility rate, the age-specific fertility rate, and the total fertility rate. The general fertility rate is the number of live births per 1,000 women of childbearing age (15 to 49 years old) per year. The age-specific fertility rate is the number of live births per 1,000 women of a specific age group per year, such as girls aged 15 to 20. The total fertility rate is the average number of children a woman will bear during her life, based on the current age-specific fertility rate.

5. The demographic statistics used to access quality of life are population density, urbanization, life expectancy, infant mortality rate, and childhood mortality rate.

Chapter 9

True/False

1. F
2. T
3. F
4. F
5. T

Fill in the Blank

1. demographic transition
2. population policy
3. family planning
4. preconception
5. demographic trap

Multiple Choice

1. C
2. C
3. C
4. A
5. A

Short Answer

1. The Mexico City policy was the U.S. government's policy of prohibiting funding for organizations involved in abortion-related activities or for countries where family planning activities were deemed coercive.

2. Family planning is a term used to describe a wide variety of measures that enable parents to control the number of children they have and the spacing of their children's births. These measures include education in hygiene, human sexuality, prenatal and postnatal care, and birth control.

3. In the U.S., the two groups least likely to use birth control are teenagers and poor women.

4. The three factors that contributed to lower population growth in the industrialized world around 1900 were (1) the rise in living standards due to the Industrial Revolution, (2) access to safe and inexpensive means of birth control, and (3) the increased cost of child rearing.

5. The four stages of the demographic transition are as follows:

   Stage 1: Birth and death rates are both high.

   Stage 2: Death rates fall but birth rates remain high (period of rapid growth).

   Stage 3: Birth rates begin to fall; growth rate approaches zero.

   Stage 4: Population growth rate continues to decline to zero or a negative rate.

Chapter 10

True/False

1. F
2. T
3. F
4. T
5. F

Fill in the Blank

1. carbohydrates, fats
2. fish
3. malabsorptive hunger
4. land races
5. cryopreservation

Multiple Choice

1. B
2. A
3. D
4. A
5. C

Short Answer

1. Proteins are important for a healthy diet because they form muscles, organs, antibodies, and enzymes.

2. The top four crops in the world are wheat, rice, maize (corn), and potato.

3. Famine is widespread starvation.

4. The "great hunger belt" is the equatorial region of various nations in southeast Asia, the Indian subcontinent, the Middle East, Africa, and Latin America where most of the world's hungry live.

5. A statistical rating of living conditions in various countries, the International Human Suffering Index tells us what conditions are like with respect to a nation's economy, health and nutrition, education, communications, and governance. Where suffering is the greatest, people tend to be poor, have little or no health care, and have little or no education. Economic conditions are bleak; population growth rates typically are high; and political power is concentrated in the hands of a few, so human rights generally are ignored or abused.

Chapter 11

True/False

1. F
2. T
3. T
4. F
5. F

Fill in the Blank

1. nonrenewable resources
2. renewable
3. energy efficiency
4. life-cycle cost
5. "drain America first"

Multiple Choice

1. D
2. C

3. B
4. D
5. B

Short Answer

1. Lightbulbs get hot because their net efficiency is only about five percent, that is, about 95 percent of the electrical energy is converted to heat.

2. Life-cycle cost is the sum of a product's purchase cost plus the operating costs incurred over the lifetime of the product (life-cycle cost = initial cost + lifetime operating cost).

3. Energy efficiency is the percentage of total energy input that does useful work and is not converted into low quality heat.

4. An energy policy is a nation's overall approach to energy resources, including both the kind and amount of energy used.

5. Also known as the law of conservation of energy, the first law of thermodynamics states that energy can be neither created nor destroyed, but it can be changed or converted in form.

Chapter 12

True/False

1. F
2. F
3. F
4. T
5. T

Fill in the Blank

1. 300, Carboniferous
2. sunlight, photosynthesis
3. proven (economic) resources
4. anthracite
5. demand side management

Multiple Choice

1. C
2. B
3. C
4. D
5. A

Short Answer

1. Coal is formed when dead plants sink to the bottom of wetlands and are compressed by increasing layers of water and sediments.

2. Petroleum is formed when aquatic organisms (algae and plankton) sink to the bottom of shallow, nutrient-rich seas. Heat from the Earth's interior and pressure from overlying sediments produce the conditions for the formation of oil.

3. Natural gas is formed when all types of organic matter are subjected to increasing amounts of heat and pressure and accelerated rates of anaerobic decomposition.

4. About 3,000 products are created from petroleum; gasoline is the most heavily consumed petroleum-based product in the United States.

5. Coke is coal that has been heated in an airtight oven; it is burned with iron ore and limestone to produce pure iron for steel.

Chapter 13

True/False

1. F
2. F
3. T
4. T
5. T

Fill in the Blank

1. uranium-235
2. convection, conduction, radiation
3. 13 m.p.h.
4. dry steam, wet steam, hot water
5. bagasse

Multiple Choice

1. B
2. B
3. D
4. B
5. A

Short Answer

1. When an atom is bombarded by free neutrons, it undergoes fission — it splits and produces smaller atoms, more free neutrons, and heat.

2. Deuterium oxide, or heavy water, is used as a coolant in heavy water reactors because it does not absorb neutrons readily.

3. A power tower is a central receiving system, usually a tall tower, in which gas or fluid is heated by the sun to power a turbine.

4. Trash conversion is the use of solid waste to produce energy, often by incineration or methane recovery.

5. Some advantages to biomass include its ready accessibility and fairly inexpensive cost. Biomass also is a relatively clean fuel, as the carbon dioxide released by burning plant matter can be offset by replanting.

Chapter 14

True/False

1. F
2. F
3. T
4. F
5. T

Fill in the Blank

1. greenhouse gases
2. acid surges
3. photochemical smog
4. rain shadow effect
5. thermosphere, mesosphere, stratosphere, troposphere

Multiple Choice

1. D
2. B
3. C
4. A
5. B

Short Answer

1. The components of clean, dry air are 78 percent nitrogen, 21 percent oxygen, and one percent carbon dioxide and rare gases (helium, argon, krypton).

2. The six primary air pollutants are carbon dioxide (CO_2), carbon monoxide (CO), sulfur oxides (SOx), nitrous oxides (NOx), hydrocarbons, and particulates.

3. The major indoor air pollutants are formaldehyde, radon 222, tobacco smoke, asbestos, combustion products of gas stoves and furnaces (CO, NOx, SO_2 and particulates), pesticides and household chemicals, and disease-causing organisms or spores.

4. Temperature inversions are caused by radiational cooling, downslope air movements, and high pressure cells.

5. The largest unregulated source of air pollution in the U.S. is wood burning stoves.

Chapter 15

True/False

1. T
2. F
3. T
4. T
5. T

Fill in the Blank

1. oligotrophic
2. epilimnion, thermocline, hypolimnion
3. littoral, euphotic, neritic, pelagic, abyssal
4. biological oxygen demand (BOD)
5. sediments

Multiple Choice

1. D
2. C
3. C
4. A
5. A

Short Answer

1. A watershed is the entire runoff area of a particular body of water.
2. Defining characteristics of eutrophic lakes include warm water, sandy or muddy bottoms, low dissolved oxygen content, and high productivity.
3. In an unconfined aquifer, the material above the water table is permeable; in a confined aquifer, an impermeable rock or clay layer prevents or restricts water flow to the water table.
4. Effluent refers to the water that leaves a wastewater treatment facility.
5. Chlorine, quicklime, and fluoride are commonly added to drinking water supplies. Chlorine kills bacteria; quicklime reduces acidity and prevents corrosion; and fluoride prevents tooth decay.

Chapter 16

True/False

1. F
2. T
3. F
4. T
5. T

Fill in the Blank

1. fertility
2. soil structure (tilth)
3. A horizon (topsoil)
4. parent material
5. soil loss tolerance level (T-value, replacement level)

Multiple Choice

1. B
2. B
3. D
4. B
5. A

Short Answer

1. On average, the abiotic composition of soil includes 45 percent minerals, 25 percent water, 25 percent air, and five percent humus.

2. The three soil categories based on texture are loams, clays, and sands. Loams — containing 40 percent silt, 40 percent sand, and 20 percent clay — have the best texture for growing most crops. Clays retain large amounts of water, leading to easily compactable soil. Sands are highly porous and retain little moisture.

3. The five interacting factors that form soil are (1) parent material, (2) climate, (3) topography, (4) living organisms, and (5) time.

4. The two major causes of agricultural soil erosion are the cultivation of marginal cropland and the use of poor farming techniques on good cropland.

5. A perennial polyculture is the cultivation of a mixture of self-sustaining (perennial) crops that do not require new planting each year; thus, these crops maintain their root systems and reduce agricultural soil erosion.

Chapter 17

True/False

1. T
2. F
3. F
4. T
5. F

Fill in the Blank

1. genetic diversity
2. genetic erosion
3. hybrids

4. charismatic megafauna
5. coral reefs

Multiple Choice

1. D
2. A
3. C
4. D
5. B

Short Answer

1. The total number of species on Earth is unknown. Estimates range from three to 100 million; the figure is probably closer to the high end of this range, with a large number of microbe and insect species.

2. Tropical rain forests and coral reefs are the two types of ecosystems with the greatest species diversity.

3. Charismatic megafauna are usually endangered mammals or birds that are culturally valued; their plights often receive funding and media attention. Vermin are animals that historically have been hated by some people; often, much money and time have been spent working toward the eradication of the undesired species.

4. The three ways that zoos group animals are taxonomic, climatic, and zoogeographic. In taxonomic grouping, animals of the same genus are placed together (such as cats or primates); in climatic grouping, animals are organized by biomes (such as tropical rain forest or savanna); in zoogeographic grouping, animals are categorized on the basis of the broad geographic area in which they live, with an emphasis on ecology and habitat (such as Africa or Australia).

5. The International Species Inventory System (ISIS) is a computer-based information system for wild animal species in captivity; it is used to prevent inbreeding.

Chapter 18

True/False

1. T
2. T
3. F
4. F

5. F

Fill in the Blank

1. magma
2. ore
3. overburden
4. strategic
5. stockpiles

Multiple Choice

1. B
2. B
3. A
4. B
5. D

Short Answer

1. The three types of rock are igneous, sedimentary, and metamorphic. Igneous rocks form when magma cools and solidifies. Sedimentary rocks form when sediments are compacted and cemented together under high temperature and pressure. Metamorphic rocks form when rocks lying deep below the Earth's surface are heated until their original crystal structures are lost and new structures form as the rocks cool.

2. Minerals are broadly classified as fuels or nonfuels. Nonfuel minerals are further classified as metallic or nonmetallic. Metallic minerals are subdivided into ferrous or nonferrous classifications. Metallic minerals are distinguished by their malleability, ductility, thermal conductivity, and electrical conductivity; nonmetallic minerals do not exhibit these characteristics. Ferrous minerals contain iron or elements alloyed with iron to make steel; nonferrous minerals contain metallic minerals not commonly alloyed with iron.

3. Critical minerals are essential to a nation's economic activity, whereas strategic minerals are essential to a nation's defense.

4. The Law of the Sea is a United Nations treaty that establishes exclusive economic zones that are under control of coastal nations; it also establishes an International Seabed Authority to license mining companies and collect taxes on minerals.

5. Compared to traditional materials, advanced materials use reduced amounts of substances and have more of a desired property (such as strength; hardness; or a certain thermal, electrical, or chemical property). Advanced materials also offer the

promise of energy conservation, better performance at lower prices, and reduced dependence on imports of critical and strategic minerals.

Chapter 19

True/False

1. T
2. T
3. T
4. T
5. F

Fill in the Blank

1. half-life
2. radioactive fallout
3. irradiation
4. cancers
5. decommissioned

Multiple Choice

1. D
2. B
3. A
4. B
5. C

Short Answer

1. Expected after a nuclear war, nuclear winter refers to a severe drop in global temperatures caused by soot, smoke, and debris blocking the sunlight.

2. The world's first commercial nuclear reactor was built near Pittsburgh, Pennsylvania (the Shippingport Atomic Power Station) in 1957.

3. According to the stipulations of the 1968 Non-Proliferation Treaty, the United States and the former Soviet Union would not provide nuclear weapons to other countries, nor would they assist other countries in developing them. The two nations also agreed to facilitate the development of peaceful uses of nuclear energy.

4. A meltdown occurs when the reactor core of a nuclear power plant becomes so hot that the fuel rods melt, burning through the containment vessel and boring into the Earth.

5. "Not In My Backyard" (NIMBY) is the mentality that develops when people fear radioactive waste may be placed in their neighborhood.

Chapter 20

True/False

1. T
2. T
3. F
4. F
5. T

Fill in the Blank

1. teratogenic
2. chemical, petroleum-refining, primary metals
3. solidification
4. secure landfills
5. Superfund

Multiple Choice

1. B
2. D
3. C
4. B

Short Answer

1. Toxic substances have the potential to cause injury to living organisms. When the possibility exists that plants or animals will be exposed to toxic substances, they are considered hazardous substances.

2. Two broad pollution prevention strategies are source reduction and waste minimization.

3. Techniques used in secure landfills to prevent or control rainwater percolating through wastes include a clay or chalk layer between the landfill and aquifer, a

waterproof plastic lining on the sides and bottom (and sometimes the top) of the landfill, a leachate collection system, and a clay cap.

4. A waste exchange is a system that brings together companies that have waste and companies that want to use it.

5. The four hazardous waste characteristics defined by RCRA are ignitability, corrosivity, reactivity, and toxicity.

Chapter 21

True/False

1. F
2. F
3. T
4. T
5. F

Fill in the Blank

1. agriculture and mining
2. 66 percent
3. tipping fee
4. refuse-derived fuel incinerator
5. precycling

Multiple Choice

1. A
2. B
3. C
4. B
5. C

Short Answer

1. A secure landfill is designed to contain hazardous wastes. A sanitary landfill is designed to receive non-hazardous residential, commercial, and industrial waste.

2. The three types of plastics that are most commonly recycled are PET, HDPE, and polystyrene.

3. The three types of soil-dwelling bacteria responsible for most of the decomposition that occurs in landfills are cellulolytic bacteria, acidogens, and methanogens. Cellulolytic bacteria separate the cellulose (a complex sugar) from wood and paper; acidogens ferment the cellulose into weak acids; and methanogens convert the acids into carbon dioxide and methane.

4. A "biodegradable plastic" is a plastic that is mixed with a biodegradable substance such as cornstarch. However, only the cornstarch biodegrades; the plastic remains chemically unchanged.

5. Green marketing in the practice of promoting products based on claims that they help or are benign to the environment.

Chapter 22

True/False

1. F
2. T
3. T
4. F
5. F

Fill in the Blank

1. National Resource Lands
2. nonconsumptive, recreational use
3. 1906 Antiquities Act
4. National Parks
5. below-cost timber sales

Multiple Choice

1. D
2. C
3. B
4. B
5. B

Short Answer

1. Some of the problems plaguing wildlife refuges include harmful secondary activities (such as the excavation of fossil fuel resources) and activities on adjacent lands (such as irrigation runoff, encroaching development, wastewater discharges, soil erosion, and toxic dump contamination).

2. A split estate is a situation in which the government owns land, but private citizens or companies own the minerals beneath its surface.

3. A duck stamp is a stamp that hunters are required to purchase each year; proceeds are used for the acquisition and maintenance of refuges.

4. The five land uses the National Forest Service is mandated to balance are (1) outdoor recreation, (2) range, (3) timber, (4) watershed, and (5) wildlife and fish habitat.

5. The 1916 Organic Act formed the National Park Service.

Chapter 23

True/False

1. F
2. T
3. T
4. F
5. T

Fill in the Blank

1. Wilderness Preservation Act of 1964
2. Wilderness Society
3. wilderness study area (WSA)
4. de facto wilderness
5. chaining

Multiple Choice

1. B
2. C
3. A
4. C
5. D

Short Answer

1. According to Robert Marshall, the most important attributes of a wilderness area are that it requires any individual who exists in it to depend exclusively on personal efforts for survival; it preserves as nearly as possible the primitive environment (and thus bars roads, power transportation, and settlement); and it is of a large size.

2. Howard Zahniser was a former president of the Wilderness Society and author of the Wilderness Act of 1964.

3. The five objectives of wilderness management under the Wilderness Act of 1964 are (1) to perpetuate long-lasting, high-quality wilderness for future generations; (2) to provide opportunities for public use and enjoyment; (3) to allow indigenous wildlife to develop through natural processes; (4) to maintain watersheds and airsheds in a healthy condition; and (5) to maintain the primitive character of wilderness as a benchmark for ecological studies.

4. Roadless Area Review and Evaluation (RARE) was conducted by the Forest Service in 1972 in order to recommend areas for wilderness designation. In 1977, RARE II offered another set of recommendations.

5. Wild lands of sufficient size can preserve ecosystem diversity, species diversity, and genetic diversity. They also protect airsheds and watersheds and are valuable for many research purposes.

Chapter 24

True/False

1. F
2. T
3. T
4. T
5. F

Fill in the Blank

1. historic preservation
2. nonmaterial (living)
3. acculturation
4. intrinsic
5. National Register of Historic Places

Multiple Choice

1. B
2. A
3. D
4. C
5. D

Short Answer

1. Historic preservation is concerned strictly with material culture (such as buildings, art, and tools). Cultural resources management preserves both material and nonmaterial culture (such as folklore, customs, and language).

2. Acid precipitation threatens material culture resources by weakening and corroding marble and limestone. Combined with ultraviolet light, acid precipitation weakens cellulose fibers in wood, thus enabling the UV light to decompose the lignin that holds the cellulose fibers together.

3. Many park units also are places of historical significance for various ethnic communities. By including members of these communities in the operation and preservation of park units, the National Park Service can relay a more accurate account of the area's cultural importance.

4. United Nations Educational, Scientific, and Cultural Organization (UNESCO) is an international agency whose purpose is to protect the global cultural heritage.

5. The World Heritage List is a list of cultural and natural properties of universal value, as determined by UNESCO'S World Heritage Committee.

Chapter 25

True/False

1. T
2. F
3. T
4. T
5. F

Fill in the Blank

1. land
2. immanence

3. Amish
4. frontier
5. Pacific yew tree

Multiple Choice

1. C
2. B
3. C
4. C
5. A

Short Answer

1. Besides religious tenets, factors that play a role in determining people's behavior toward the environment include personal interpretation of religious teachings, how strictly one adheres to his or her beliefs, and personal ethic and morals.

2. Floresta is a Christian organization founded in 1984 to address economic problems in developing countries that cause and are caused by deforestation.

3. Three traditions of Earth-based spiritualities are (1) celebration of the cycle of life; (2) immanence, the divine embodied in the living world and its components; and (3) interconnection, the relatedness of al things.

4. An ethic is a system or code of morals that governs attitudes and behaviors.

Chapter 26

True/False

1. T
2. T
3. T
4. F
5. F

Fill in the Blank

1. steady state economy
2. culture of maintenance
3. nonrenewable resources

4. research expeditions

5. debt-for-nature swap

Multiple Choice

1. C

2. A

3. C

4. D

5. C

Short Answer

1. Green taxes, also called environmental taxes, are fees added to the price of products or practices that have high environmental costs.

2. Sustainable resource use is the use of renewable resources at rates which do not exceed their capacity for renewal. Sustainable development improves the quality of human life while living within the carrying capacity of supporting ecosystems. Sustainable development is possible only when resources are used in a sustainable manner.

3. Cost-benefit analysis is a comparison of the estimated costs (losses) of a proposed project with the estimated benefits (gains). Cost-benefit analysis makes it possible to weigh the worthiness of a project in a straightforward manner; however, to be accurate, cost-benefit analysis must include external costs (environmental and social), which often are difficult to measure.

4. Sections of Antarctica are claimed by seven nations — Australia, New Zealand, South Africa, Norway, Argentina, Chile, and the United Kingdom. Many nations argue for the right to claim Antarctica's vast resources; however, research is the only industry permitted on the continent.

5. In a traditional economy, individuals and the community as a whole make all decisions about what foods to grow and what items to produce in order to survive; they are self-reliant and self-sufficient. In a pure command economy, the government makes all decisions about what to produce, what price to charge, and where and how to distribute goods. In a pure market economy, economic decisions are made by buyers and sellers in the marketplace — not the government — and are based on interactions of supply, demand, and price.

Chapter 27

True/False

1. F
2. F
3. F
4. T
5. F

Fill in the Blank

1. litigation, dispute resolution
2. Environmental Protection Agency
3. nuisance
4. mediator
5. common law

Multiple Choice

1. D
2. C
3. D
4. D

Short Answer

1. An abatement is a court order that requires the defendant to stop or restrict a specified act; it is an example of an injunction.

2. The three causes of action upon which cases involving common law are based are nuisance, trespass, and negligence.

3. Nuisance is an unwarranted, unreasonable or unlawful use of a person's own property that produces annoyance, inconvenience or material injury to another (for example, smoke blowing from one person's burning field onto another's property).

4. Negligence is the failure to exercise the care that "a prudent person" usually takes, resulting in an action or inaction that causes personal or property damage (for example, designing a pipe in such a way that it eventually will leak gas into an underground aquifer).

5. Coordinated resource management planning (CRMP) is a dispute resolution strategy that brings together a team of resource managers and users to develop resource management plans.

Chapter 28

True/False

1. F
2. F
3. T
4. T
5. F

Fill in the Blank

1. environmental education
2. Institute for Earth Education
3. English, Driver's Education, Typing, Physics
4. Wisconsin, Maryland

Multiple Choice

1. B
2. D
3. A
4. C

Short Answer

1. Two approaches to adding environmental education to a school curriculum are as follows:

    (1) Non-integrated environmental science courses — Classes in which environmental information is presented to students by teachers trained in such areas. Other classes and teachers are unaffected by this addition to the school's curriculum.

    (2) Integrated environmental issues — A method whereby teachers integrate environmental issues into all classes in the curriculum. Environmental topics are studied as they are appropriate to each class.

2. Target groups for informal environmental education include elected officials, business leaders, and members of the media. It is important that these groups become environmental aware because their action or inaction can influence local environmental quality.

3. The Roots and Shoots program deals with the following three project areas: (1) caring for animals, (2) caring for the environment, and (3) caring for the human community.

4. Five forums for informal environmental education are (1) zoos, (2) nature centers, (3) museums, (4) aquariums, and (5) wildlife refuges.